Differential Equations
With Derive ®

Differential Equations
With Derive ®

David C. Arney
United States Military Academy

MathWare
604 E. Mumford Dr.
Urbana, Il.

Limits of Liability and
Disclaimer of Warranty

The author has written this book to show the varied uses of the program Derive in learning math. The author and publisher make no warranty of any kind, expressed or implied, with regard to the program or documentation contained in this book. The author and publisher shall not be liable in any event for incidental or consequential damages in connection with, or arising out of, the furnishing, performance, or use of the program.

This text was not prepared as part of official duties and is not endorsed by the Army or the Department of Defense.

DERIVE® is a registered trademark of Soft Warehouse Inc. Honolulu, HI.

Contents

Preface

For it is unworthy of excellent men to lose hours like slaves in the labor of computation.

—Gottfried Wilhelm Leibniz

This book is designed to show how to use the software package Derive®[1] to help solve problems from differential equations and related subjects. It is a companion to any of the textbooks used in differential-equations courses or related-subject courses (i.e., engineering mathematics, applied mathematics, dynamical systems).

Derive is a computer algebra system (CAS) that provides capabilities to perform symbolic, graphic, and numeric manipulations. It is mostly menu driven, making it user-friendly, which is necessary for an educational problem-solving tool. Its screen displays are nicely formatted and easy to read. In capable hands, Derive is a powerful tool that can help solve problems encountered in differential-equations courses and in other subjects as well.

This manual is organized into seven chapters. The first chapter contains a general explanation of the basic capabilities and limitations of Derive. This section generally covers the symbolic, graphic, and numeric capabilities of the

---

[1]Derive® is a registered trademark of Soft Warehouse, Inc., Honolulu, HI.

software in the areas of algebra, calculus, matrix algebra, complex variables, and special functions. The basic keystrokes, menu commands, mode selection, plotting parameters, and screen displays are presented, along with examples on the use of Derive commands to perform simple operations in algebra, trigonometry, and calculus.

The second through seventh chapters contain examples and exercises similar to those found in differential-equations textbooks. The examples show how to use Derive to perform some of the manipulations, plotting, and analysis to solve the problem. Some of the problems involve differential-equation models from applications. Other problems deal directly with the mathematical concepts of topics in the course. The laboratory exercises require the reader to use Derive to solve, explore, and analyze problems. Many of these problems are similar to those provided in the examples. Other problems introduce new concepts and extensions of both the mathematics and the software. These laboratory exercises lead the reader through the steps of problem solving. Sometimes the exercises involve conducting a computational experiment to determine an answer. Students should perform these laboratory exercises and experiments to develop their problem-solving skills and Derive keystroking skills.

The second chapter covers the basic tools of calculus: graphing, differentiation, and integration. The third chapter presents first-order differential equations. The fourth chapter presents numerical methods and difference equations. The fifth chapter covers techniques for solving second-order differential equations. The sixth chapter covers both matrix algebra and systems of differential equations. Partial differential equations is the topic presented in the last chapter.

In writing this manual, I have assumed that the reader either is already familiar with differential equations or is concurrently studying the subject. The object of this book is not to teach differential equations, but to help in that endeavor. I refer students who are studying calculus and instructors of calculus courses to a similar manual about using Derive in calculus: *Exploring Calculus with Derive*, published by Addison-Wesley.

It is hoped that this guide can contribute to better student understanding and performance in problem solving. It illustrates the power of Derive as a computational tool that enables students to become better problem-solvers.

Many people have helped make this book possible. I give special thanks to Frank Giordano for his encouragement. I thank the creators of Derive, David Stoutemyer and Albert Rich, for their excellent software and suggestions. Finally, I give special thanks to my family for their support and understanding.

Chris Arney
West Point, NY
17 November 1992

Differential Equations
With Derive®

Notation and Technical Information

"What is the use of a book," thought Alice, *"without pictures or conversations?"*

—Lewis Carroll, *Alice in Wonderland* [1865]

This manual was typeset using LaTeX. This text processing system allows for special notation and fonts to be used to help identify the computer keystrokes and input commands. Boldface type is used to indicate the precise key on the keyboard to strike. This is especially helpful for identifying the special keys like **Enter**, **Ctrl**, **Alt**, and the function keys (**F1**, **F2**, ..., **F10**). The **Ctrl**, **Alt**, and **Shift** keys are often used in combination with another key, and this is denoted by placing a hyphen between them and showing both keys in boldface (i.e., **Alt-e** or **Ctrl-Enter**).

The teletype font is used for the functions and menu selections for input to or output from the software (for example, `Author`, `Simplify`, `Help`, `Calculus`, `LN(2z/3)` , `cos(xy-x)` , `DIF(e^x+sin(x),x,2)`). When two or more menu commands are given in sequence, they are separated by a space (i.e., `Declare Variable`). The in-line functions are usually in uppercase to distinguish them from the menu selections, and are displayed in a box for further emphasis when they are in the exact form to be entered into the

1

computer. Derive is not sensitive to case unless requested to be, so actual input ordinarily may be made with uppercase or lowercase characters.

Sometimes italics are used to indicate a pseudo-command. For example, the selection of the Author menu command and the input of an expression can be denoted by *Authoring* the expression. This notation is not used often and only after the command is very familiar to the reader. Italics are also used when key elements of Derive are introduced and defined.

Words in uppercase letters and in regular typeface are used to indicate DOS file names. Derive uses utility files to store extra commands and functions, system states, and output expressions. These files are loaded whenever their commands are needed.

Version 2.5 of Derive is used in all the examples in this book. Earlier versions of the software contained utility files with different functions, which play a significant role in solving many problems. See Section 1.15 for more information about other versions of Derive.

There are numerous figures throughout the manual showing actual screen images from Derive. Usually, the highlight is moved off the visual region of the screen so it will not interfere with the display. The screen images were produced using the CAPTURE program of Microsoft Word and plotted in Postscript format on an NEC laser printer. The computer used to run the software was a Zenith 248Z, which is compatible with an IBM-AT.

1

Getting Started

Today we render unto the computer what is the computer's, and unto analysis what is analysis'; we can think in terms of general principle, and appraise methods in terms of how they work . . . .

—Peter Lax [1989]

This chapter describes the fundamentals of using Derive to solve mathematics problems and is a good preparation for the example problems solved in the remaining chapters. Since differential equations involve many areas of mathematics, sections are included on the use of Derive in the subject areas of calculus, matrix algebra, and complex variables. This chapter also includes the fundamental commands of the software in symbolic algebra, plotting, and numerical approximations. One section demonstrates the use of Derive commands to solve simple drill problems in algebra, trigonometry, calculus, and differential equations. However, neither this manual nor this first chapter are complete reference manuals for Derive. This book is not intended to replace any of the user manuals that come with the software. In fact, this manual will refer you at times to the *Derive User Manual* for more information. Another section in this chapter compares the software versions of Derive.

1.1   An Overview of Derive

> *Making mathematics more exciting and enjoyable is the driving force behind the development of Derive. The system is destined to eliminate the drudgery of performing long tedious mathematical calculations.*
>
> —*Derive User Manual* [1990]

It is important to realize that the Derive software was not specially designed to solve problems in just one field, such as differential equations. Derive is a computer algebra system (CAS) whose function is more general than any one subject or topic. It has capabilities to perform symbolic, numeric, and graphic operations. There are other packages like Phaser[1], MDEP[2], MacMath[3], and Differential Equations Graphics Package[4] that are designed for the express purpose of analyzing and numerically solving differential equations. Software programs like LinTek[5], MAX[6], MATLAB[7], LINDO[8], and the Linear Algebra Toolkit[9] help solve matrix and linear algebra problems. All these packages possess some of the capabilities of a CAS, but usually lack the versatility of CAS software.

Derive performs many of the mathematical steps and approximations involving algebra, calculus, trigonometry, number theory, and plotting necessary to solve, analyze, and study many types of problems. Other CAS software packages similar to Derive for personal computers are Maple[10], Mathematica[11], muMath[12], Macsyma[13], Reduce[14], Mathematics Exploration Toolkit[15], and Theorist[16].

---

[1]H. Kocak, Springer-Verlag.
[2]J.L. Buchanan, United States Naval Academy.
[3]J.H. Hubbard and B.H. West, MacMath.
[4]Sheldon Gordon, MatheGraphics.
[5]John Fraleigh, Addison-Wesley.
[6]E.A. Herman and C.H. Jepsen, Brooks-Cole.
[7]Little and Moler, The Math Works.
[8]Scientific Press.
[9]C. Wilde, Addison-Wesley.
[10]Waterloo Maple Software.
[11]Wolfram Research.
[12]The predecessor of Derive.
[13]Symbolics.
[14]Northwest Computer Algorithms.
[15]WICAT Systems.
[16]Prescience.

The following figure shows part of a sample Derive screen display from a problem-solving session showing the general layout of Derive's user interface:

$$1: \quad 4x^5 - 16x^4 - 8.5x^3 + \sqrt{5}x^2 - x + 15$$

$$2: \quad \frac{d}{dt}\left[8^{2t} - \frac{LN(t)}{t^2 - 1} \right]$$

$$3: \quad \int_0^{\pi} \frac{SIN(4x-3)}{x}\,dx$$

COMMAND: Author Build Calculus Declare Expand Factor Help Jump soLve Manage
          Options Plot Quit Remove Simplify Transfer moVe Window approX
Enter option
User   Free:100%   Derive Algebra

Derive screen showing three working expressions in the work area
(expression # 3 is highlighted) and the main menu.

The screen is organized into several sections, each with an important function. The top part contains the working expressions. This sample of this section of the Derive screen shows expressions 1–3.

$$1: \quad 4x^5 - 16x^4 - 8.5x^3 + \sqrt{5}x^2 - x + 15$$

$$2: \quad \frac{d}{dt}\left[8^{2t} - \frac{LN(t)}{t^2 - 1} \right]$$

$$3: \quad \int_0^{\pi} \frac{SIN(4x-3)}{x}\,dx$$

Three expressions (numbered 1–3) in the top part of Derive screen.

This section can also include plots and can be partitioned into subsections by the user. The screen below shows three windows, one containing algebraic expressions, another window containing a 2-dimensional plot, and the third has a 3-dimensional plot. Windowing and plotting are discussed in Sections 1.4 and 1.10, respectively.

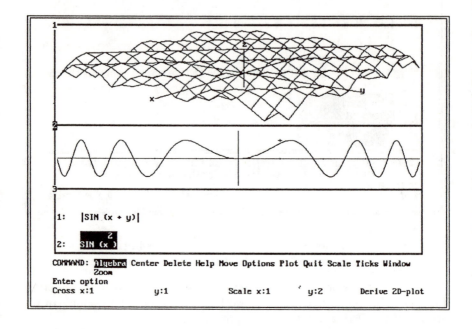

Derive screen with a 3-dimensional plot, a 2-dimensional plot, and algebraic expressions.

1.2   Menus, Commands, Functions, and Utility Files

Computers change not only how mathematics is practiced, but also how mathematicians think.

—Lynn Arthur Steen [1990]

The section below the expressions on the Derive screen contains the menu. *Commands* are selected from the menu in one of two ways. The first way is merely to type the letter associated with the menu command (the uppercase letter in the display of the command, which is not always the first letter of the word). The second way is to move the highlight in the command line using the **Space**, **Tab**, **Backspace**, or **Shift-Tab** keys and pressing the **Enter** key when the desired menu command is highlighted.

Functions are typed directly into the work area using the `Author` menu command. *Utility files* contain additional functions. Utility files can be loaded whenever their functions are needed. The **Esc** key is used to abort a command or menu without making any changes.

The main menu and the `Author` command are the user's usual interfaces with Derive, although the `Simplify` and `approX` commands are also used frequently.

```
COMMAND: Author Build Calculus Declare Expand Factor Help Jump soLve Manage
         Options Plot Quit Remove Simplify Transfer moVe Window approX
Enter option
                                    Free:100%              Derive Algebra
```

Derive's familiar main menu and its command selections.

Many of the commands are executed directly from the main menu. These executable commands include `Author`, `Build`, `Expand`, `Factor`, `Help`, `Jump`, `soLve`, `Quit`, `Remove`, `Simplify`, `approX`, and `moVe`. The following table gives a short description of each of these executable main menu commands:

Command	Function
Author	Allows for the input of expressions into the algebra window or work area.
Build	Provides for the building of an expression from previous expressions.
Expand	Performs the algebraic expansion of an expression.
Factor	Factors an expression. If the expression is a polynomial, factors it into the lowest degree terms meeting the criterion established with the mode portion of the command. If the expression is a number, expresses the number in terms of its prime factors.
Help	Provides help on the syntax and function of commands (see Section 1.5).
Jump	Moves the highlight to a given expression number. This is a good way to move to an earlier expression in the work area.
soLve	Solves an equation for the desired variable.
Quit	Stops the Derive program. Control returns to DOS.
Remove	Removes expressions from the work area.
Simplify	Simplifies an expression, equation, or a Derive command in the work area. This command is used frequently.
moVe	Rearranges expressions in the work area.
approX	Approximates the value of an expression. Precision is set with the Options Precision command.

Main menu commands and their functions.

Some other commands merely open submenus that contain additional commands. The submenu commands also can either execute a command or open a submenu. Therefore, the menu structure is much like a tree with several branches. The user must learn what branch to follow to obtain the needed executable command, or learn the proper command name so it can be typed directly into the work area. Some commands can be entered both ways, while others are available only through the menu or only through a typed command.

The following figure shows a schematic of the tree-like menu structure. Only the most common commands are included in the figure.

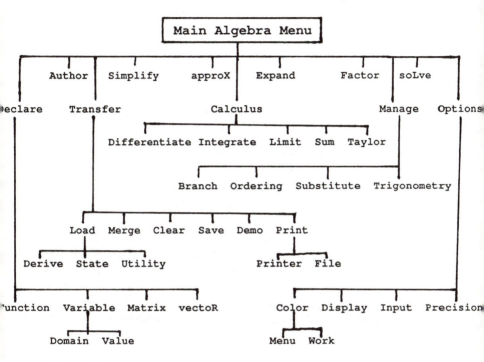

Most of the important submenus and their associated commands are explained in the following paragraphs and tables. Other submenus are discussed in later sections of the manual.

One such submenu selection that will be very useful for our purposes is Calculus. This submenu's commands are Differentiation, Integration, Limit, Product, Sum, and Taylor. These calculus menu commands also have equivalent functions. These commands and functions are described in more detail in Section 1.7.

The set of commands in the submenu Declare are Function, Variable, Matrix, and vectoR. These are very useful and powerful commands, and careful thought in their use is often needed to solve problems correctly. These commands will be used in some of the examples and may be necessary in some of the laboratory exercises. Not only can you declare these mathematical structures (matrix, vector, etc.), but you can also establish important domain information on variables using the Declare Variable Domain command, and the components in the structures can be directly entered. The following table describes the function of these commands:

Command	Function
Function	Declares a function and optionally defines it.
Variable	Assigns the domain of a variable (options include Positive, Nonnegative, Real, Complex, and Interval) or assigns the variable a value.
Matrix	Establishes the dimensions of a matrix and provides for the entry of elements in the matrix.
vectoR	Establishes the dimension of a vector and provides for the entry of elements in the vector.

Declare menu commands and their functions.

The Manage submenu contains the commands Branch, Exponential, Logarithm, Ordering, Substitute, and Trigonometry. The most important and frequently used command in this list is probably the Manage Substitute command, which substitutes values for variables or subexpressions of an expression. The location of the Substitute command in the Manage submenu is worth remembering. More information on these commands is found in the *Derive User Manual*.

The Options submenu sets up different modes of operation for Derive. Some of the Options commands are usually used before any evaluations are done. Other commands may be performed during the course of problem solving. The commands in this submenu are Color, Display, Execute, Input, Mute, Notation, Precision, and Radix. The following table provides a short description of these commands. It is important to set the proper parameters for your computer using the Options Display command. These settings can be saved via the Transfer Save State command.

Command	Function
Color	Changes the color of the work area, menu, or plotting curves.
Display	Sets Mode to Text or Graphics and establishes the Resolution and Adapter.
Execute	Allows for the execution of DOS commands.
Input	Establishes the input mode for variables as single letters or words.
Mute	Sets the error beep on or off.
Notation	Sets the style of notation for numerical output and the number of digits displayed.
Precision	Sets the computational precision mode and the digits of accuracy for the Approximate mode.
Radix	Sets the radix base for number input and output.

Options menu commands and their functions.

The **Transfer** submenu provides for input and output in Derive. The subcommands and their functions are given in the following table. This menu is used to load utility files, to save files, to run demonstration files, to save and to load system-state files, and to obtain hard-copy printouts of the expressions in the work area.

Command	Function
Merge	Adds the expressions stored in a specified .MTH file to those in the work area.
Clear	Deletes all the expressions from the work area.
Demo	Loads and simplifies expressions from an .MTH file, one expression at a time.
Load	Loads all the expressions in a utility file, user-defined Derive (.MTH) file, or State (.INI) file into the working area or to set up Derive parameters. Also loads numeric data from data files.
Save	Saves the expressions in the working area to a file. Format options include Derive (.MTH), FORTRAN, C, Pascal, or Basic, or to save the current State. The State option saves the current state of Derive parameters to an .INI file for reloading.
Print	Prints the expressions in the work area on a printer or to a file. Allows for the setting of several printing options.

Transfer menu commands and their functions.

There are some operations in the form of functions that are not reached through any menu. These functions must be typed into the workspace using the **Author** command. Some of the more useful and common functions are CURL, FIT, GRAD, LAPLACIAN, VECTOR_POTENTIAL, STDEV, RMS, RANDOM, MOD, FLOOR, GCD, LCM, ROW_REDUCE, EIGENVALUES, DET, DIMENSION, PRODUCT, POTENTIAL, PHASE, IM, and REAL. Explanations of these operations are found in later sections of this chapter or the *Derive User Manual*, and are available on-line through the **Help** menu. Some of these operations will be used in the example problems of chapters 2 through 7.

Additional functions are available in the utility files. The following table provides a listing of the utility files along with a description of the types of operations by the functions contained in the files:

Utility File	Description
SOLVE	Functions to solve systems of nonlinear equations
VECTOR	Additional vector and matrix operations
NUMERIC	Numerical differentiation and integration
DIF_APPS	Applications of differentiation
INT_APPS	Applications of integration
ODE1	Functions to solve first-order differential equations
ODE2	Functions to solve second-order differential equations
ODE_APPR	Functions to approximate solutions to ODEs
RECUREQN	Functions to solve difference equations
BESSEL	Functions for Bessel and Airy functions
GRAPHICS	Utility functions for plotting space curves
MISC	Functions for number theory and other miscellaneous subjects

Utility files and descriptions of their use.

1.3   Keystroking and Entering Expressions

Derive not only has to be a tireless, powerful, and knowledgeable mathematical assistant, it must be an easy, natural, and convenient tool.

—*Derive User Manual* [1990]

As discussed in Section 1.2, the keystrokes to select menu commands in Derive are straightforward. Just move the highlight by hitting the space bar and press **Enter** when the desired command is highlighted, or merely type the letter that is capitalized in the desired command. However, there are many other keystrokes to learn in order to enter and edit expressions when using the Author and other similar commands. Keyboard skill is essential and a prerequisite for efficient use of the software. Learning to use the editing features can reduce considerably the amount of retyping of expressions.

One feature is the ability to input comments into the work area between expressions. Comments are entered by putting quotes around the line of input when entering an expression.

When entering and editing an expression using the Author command, the **Backspace** deletes the previous character (left of the cursor). Derive uses many of the same keystrokes as WordStar, the word-processing package. **Ctrl-S** moves the cursor one character to the left without erasing the character, and **Ctrl-D** moves the cursor one character to the right. **Ctrl-A** moves the cursor a group of characters (or token) to the left, and **Ctrl-F** moves the cursor a token to the right. **Ctrl-Q-S** moves the cursor to the left end of the line, and **Ctrl-Q-D** moves the cursor to the far right end. **Del** deletes the character at the cursor. **Ctrl-T** deletes a token at a time, and **Ctrl-Y** deletes the entire line.

Ins toggles between two typing modes: one mode inserts between existing characters, and the other mode overwrites any existing characters. When the insert typing mode is in effect, the word Insert appears on the status line below the menu.

The direction keys are used to highlight and obtain subexpressions (or entire expressions) from previous expressions in the working area. These subexpressions can be expanded or assembled into new expressions. The ↑ and ↓ keys move the highlight one expression at a time. **PgUp** and **PgDn** move the highlight several expressions at once. **Home** moves to the first expression, and **End** highlights the last expression in the working area. The other direction keys, ← and →, move the highlight through the individual tokens (or subexpressions) of an expression. Even finer highlighting is possible (see the *Derive User Manual*). The **F3** function key inserts whatever is highlighted into the author line. The **F4** key does the same and also automatically places parentheses around the expression. Similar things can be done using the Build command (see Section 1.2). The easiest way to refer to previous expressions is by their number in the work area. Just use the # symbol preceding the number of the expression to obtain that expression. For example, authoring #4/#6 inserts expression number 4 divided by expression number 6 into the work area.

The **Enter** key completes the expression and adds it to the work area. **Ctrl-Enter** also does this and automatically *Simplifies* the expression.

There are simple keystrokes to enter the common special characters. The symbolic entry for the irrational numbers π and e are **Alt-p** and **Alt-e**, respectively. π can also be entered by typing pi. The imaginary unit i is entered by **Alt-i**. The square root symbol $\sqrt{}$ is entered by **Alt-q** or by using the function SQRT—for example, SQRT(x).

There are two input modes for variable names—character or word. The mode is set with the Options Input command. In the Character mode, each letter represents a variable. In the Word mode, entire words are used as variables, and the words representing variables must be separated by a space, an operator, or appropriate punctuation, such as a parenthesis. The Options Input command also provides the mechanism to put Derive into case-sensitive (uppercase letters different from lowercase) or case-insensitive modes.

Some Greek letters can be used as variables, and entered using the **Alt** key and a letter key or by typing their Latin names. The following chart provides the keys to use to obtain Greek letters:

Greek Letter Obtained	Keystrokes for Entry of the Letter
α	**Alt-a**
β	**Alt-b**
γ	**Alt-g**
δ	**Alt-d**
ϵ	**Alt-n**
θ	**Alt-h**
μ	**Alt-m**
π	**Alt-p**
σ	**Alt-s**
τ	**Alt-t**
ϕ	**Alt-f**
ω	**Alt-o**

Keystrokes to enter Greek letters.

1.4   Screen Displays and Windows

My greatest hope is that it pleases those who have at heart the development of science and that it proposes solutions that they have been looking for or at least opens the way for new investigations.

—Carl Friedrich Gauss, *Disquisitiones Arithmeticae* [1801]

Derive allows the screen to contain more than one window. These windows can split the screen in a variety of ways or overlay one another. Windowing can be used for several purposes. Separate windows can contain different problem solutions, different developments of the same problem, or different perspectives or scales of the same plot, or they can maintain important results in one window while working in another. Windows can be especially helpful in showing symbolic computations, numeric computations, and graphics all on the same screen in three or more different windows.

The `Window` command is called from the main menu. The `Window` menu is shown here.

```
WINDOW: Close Designate Flip Goto Next Open Previous Split
Enter option
                              Free:100%           Derive Algebra
```

Windowing is a very important and powerful feature of Derive. Be sure to use windowing when it is helpful. The following table describes the use of the windowing commands:

Command	Function
Close	Closes the active window.
Designate	Designates the active window as 1 of 3 types (Algebra, 2-D plot, 3-D plot).
Flip	Flips or rotates between overlaid windows (the windows are not numbered separately).
Goto	Activates and moves control to the designated window.
Next	Activates and moves control to the next window, in sequence.
Open	Opens a new window and designates it as 1 of 3 types (Algebra, 2-D plot, 3-D plot).
Previous	Activates and moves control to the previous window, in sequence.
Split	Splits the current active window to form a new window. The split can be made vertically or horizontally and can be established in different sizes.

Window menu commands and their functions.

The following example shows three horizontal windows. The top window contains the symbolic form of the function, the middle window shows the global behavior of this function, and the bottom window gives a smaller-scale perspective of local behavior near the origin. The number on the active window is highlighted (window #3 in the following figure).

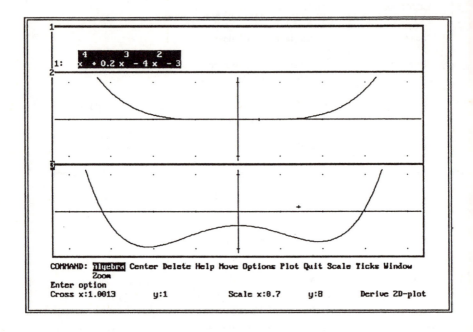

There is a convenient keystroke that saves time when moving between windows. The **F1** key is equivalent to the `Window Next` command and provides easy and rapid movement through the existing windows.

1.5   On-Line Help and System State

It is better to know some of the questions than all of the answers.

—James Thurber [1945]

The `Help` command displays the following menu:

```
                      Derive Help Menu

               E - line Editing commands
               F - Functions and constants
               A - Algebra window commands
               2 - 2D-plot window commands
               3 - 3D-plot window commands
               U - Utility file functions
               S - current State of system
               R - Return to Derive

           Press letter for desired subject






HELP: Editing Functions Algebra 2D-plot 3D-plot Utility State Resume
Enter option
                              Free:100%            Derive Algebra
```

To obtain on-screen help on any of these subjects, press the appropriate letter in this menu. Helpful information is provided along with the section number in the *Derive User Manual*, which contains more information. There is plenty of information available, so don't be hesitant to use this feature.

The `Help, S` option displays the current system state. There are several screens of information available through this command. The first screen provides the basic mode settings, and the second screen contains mostly information on the plotting parameters and output formats. Information is provided about the status of the display modes set with the `Options Display` menu commands (`Text` or `Graphics`, resolution, type of graphics adapter—CGA, EGA, VGA, etc.) by use of a numerical code. All of these modes can be saved and established in the .INI file. The system status for the set-up used to solve the examples and exercises in this manual are provided in the following output:

```
                         System Control Settings

   Algebra window numerical settings
      Precision mode:  Exact
      Precision digits:  6
      Notation style:  Rational
      Notation digits:  6
      Input radix base:  10
      Output radix base:  10
      Input mode:  Character
      Case mode:  Insensitive

   Algebra window simplification settings
      Branch selection:  Principal
      Logarithms:  Auto
      Exponentials:  Collect
      Trig functions:  Auto
      Trig powers toward:  Auto
```

```
                         System Control Settings

   Plot window plotting settings
      Rows per tick mark:  4
      Columns per tick mark:  9
      Plotting accuracy:  7
      Coordinate type:  Rectangular
      Auto change color:  No
      Axes color:  15
      Cross color:  15
      3D plots top color:  6
      3D plots bottom color:  5

   Printer settings
      Page length:  66
      Page width:  80
      Top margin:  0
      Bottom margin:  0
      Left margin:  8
      Right margin:  3
```

```
                          System Control Settings
    Window color settings
      Frame color:   15
      Option color:  15
      Prompt color:  15
      Status color:  15
      Menu background color:  0
      Border color:  0
      Work color:  15
      Work background color:  0

    Display mode settings
      Video mode:  18
      Character set:  Extended
```

These state parameters can be saved into and reloaded from an .INI file using the `Transfer Save State` and `Transfer Load State` commands, respectively. Derive starts in the state set in the file DERIVE.INI. If a different initial state is desired, save that state in the file DERIVE.INI with the `Transfer Save State` command and that will become the initial state of the software. Different states can be saved in other .INI files with the `Transfer Load State` command and loaded whenever needed. See Section 1.2 or the *Derive User Manual* for more information about the system state, and refer to Section 1.17 of this book for recommendations on establishing proper mode settings for your own DERIVE.INI file.

1.6   Symbolic Algebra

The performance of the computer is to be judged by the contribution which it will make in solving problems of new types and developing new methods.

—John von Neumann [1946]

Derive has the capability to do many algebraic manipulations using the Simplify, soLve, Manage, Expand, and Factor commands. See Section 1.2 or the *Derive User Manual* for explanation of these commands. Derive can work with equations, expressions, and inequalities. It also can do operations with ∞. Enter ∞ by typing inf.

Derive assumes the default declaration of Real for all variables, unless overridden by the Declare Variable Domain command. In a similar manner, the Declare Variable Value command creates user-defined variables and optionally gives them values. The constant number e is entered by Alt-e and displayed as ê, and the constant π is entered as pi or Alt-p. Three special files are available to the Derive user with many useful variables already declared. These files are ENGLISH.MTH, METRIC.MTH, and PHYSICAL.MTH. These files can be loaded when needed using the Transfer Load Utility commands.

Declare Function is a very powerful feature in Derive. It creates user-defined functions that can also be used as functional operators.

Some of the common algebraic and trigonometric functions in Derive are provided in the following table:

Command	Function
EXP(z)	Exponential of z, displayed as $\hat{e}^z$.
SQRT(z)	$z^{1/2}$. Takes into account the domain of the variable and branch setting.
LN(z), LOG(z,w)	Principal natural logarithm and log to the base w.
PI, DEG	Constants π and $\pi/180$.
SIN(z), COS(z), TAN(z) COT(z), SEC(z), CSC(z)	Trigonometric functions.
ASIN(z), ACOS(z), ATAN(z) ACOT(z), ASEC(z), ACSC(z)	Inverse trigonometric functions.
SINH(z), COSH(z), TANH(z) COTH(z), SECH(z), CSCH(z)	Hyperbolic trigonometric functions, which simplify to the exponential equivalents.
ASINH(z), ACOSH(z), ATANH(z) ACOTH(z), ASECH(z), ACSCH(z)	Inverse hyperbolic trig functions, which simplify to the logarithmic equivalents.
ABS(x), SIGN(x)	$|x|$ and sign of x, respectively.
MAX(x1,...,xn), MIN(x1,...,xn)	Maximum and minimum of all the arguments.
STEP(x)	1 if $x > 0$, 0 if $x < 0$. Provides for the definition of piecewise functions.

Common algebra and trigonometry functions.

The trigonometric manipulations in Derive are controlled with the command **Manage Trigonometry**. There are several possible combinations of ways to manage and simplify the trigonometric functions. Derive uses radians for angular measure in all its trigonometric functions.

There are many other functions in Derive to perform operations in probability, statistics, and finance. See the *Derive User Manual* for descriptions of these functions.

1.7   Calculus

Calculus is first of all a language, not only for scientists, but also for economists and social scientists.

—Lynn Arthur Steen [1990]

Derive performs the calculus operations of limits, differentiation, antidifferentiation, integration, Taylor polynomial approximations, series summation, and products. These operations can be done through the `Calculus` submenu or with in-line commands. Several items about these operations are worth mentioning: i) antidifferentiation does not automatically produce an arbitrary constant, ii) definite integration in the `Approximate` mode is performed with an adaptive variant of Simpson's rule, and iii) definite integrals in exact mode are evaluated using one-sided limits of an antiderivative, so it is your responsibility to account for any internal singularities.

When the `Calculus` submenu is selected, a new menu appears at the bottom of the screen. The executable commands in this submenu are as follows: `Differentiate`, `Integrate`, `Limit`, `Product`, `Sum`, and `Taylor`. The following table provides descriptions of these commands:

Command	Function
Differentiate	Finds the derivative (regular or partial) of the highlighted expression.
Integrate	Finds either the antiderivative or the definite integral of a function, depending on whether limits of integration are entered.
Limit	Finds the limit of an expression as a variable approaches a value. The approach can be from the left, right, or both.
Product	Finds the definite product or antiquotient of an expression. The index variable is required.
Sum	Finds the definite sum or antidifference of an expression. The index variable is required.
Taylor	Finds the Taylor polynomial approximation to an expression. The expansion variable, point, and degree are required.

`Calculus` menu commands and their functions.

These calculus menu commands also have equivalent in-line commands. The following table explains the format and use of these in-line commands. The in-line commands must be entered along with their arguments using the Author menu command; the Simplify command then executes the operation.

Command	Function
DIF(u,x,n)	Finds the nth-order derivative of u with respect to x.
INT(u,x)	Finds the antiderivative of u with respect to x.
INT(u,x,a,b)	Finds the definite integral of u with respect to x from a to b.
LIM(u,x,a,0)	Finds the limit of u as x approaches a from both left and right sides.
LIM(u,x,a,1)	Finds the limit of u as x approaches a from the right (above).
LIM(u,x,a,-1)	Finds the limit of u as x approaches a from the left (below).
PRODUCT(u,n)	Finds the product of u with respect to n.
PRODUCT(u,n,k,m)	Finds the definite product of u as n goes from k to m.
SUM(u,n)	Finds the sum of u with respect to n.
SUM(u,n,k,m)	Finds the definite sum of u as n goes from k to m.
TAYLOR(u,x,a,n)	Finds the Taylor polynomial approximation of order n to u about the point $x = a$.

Calculus in-line commands and their functions.

There are three utility files that contain additional calculus-related functions: DIF_APPS.MTH, INT_APPS.MTH, and MISC.MTH. See Section 1.12 for more information about these utility files.

Derive also has several vector calculus functions. The default for the vector function is the 3-dimensional rectangular coordinates x, y, and z. However, the utility file VECTOR.MTH can help perform coordinate transformation and produce operation in other coordinate systems. See the *Derive User Manual* for the description of this file and its use. The following table provides descriptions of some of the vector calculus commands:

Command	Function
GRAD(f(x,y,z))	Finds the gradient of an expression.
DIV([u(x,y,z),v,w])	Finds the divergence of a vector function.
LAPLACIAN(f(x,y,z))	Finds the divergence of the gradient (Laplacian) of a function.
CURL([u,v,w])	Computes the curl of a vector.
POTENTIAL([u,v,w])	Calculates the scalar potential of a vector.
VECTOR_POTENTIAL ([u,v,w])	Calculates the vector potential of a vector.

Common vector calculus functions.

1.8   Matrix Algebra

It will be seen that matrices comport themselves as single quantities.

—Arthur Cayley [1858]

Solving systems of linear equations and performing matrix and vector operations are important and frequent tasks in solving problems involving systems of differential equations. Derive has capabilities to help the user perform these operations.

Vectors can be entered in several ways. Four common ways are: i) `Author` an expression of the form $[x_1, x_2, \ldots, x_n]$, ii) use the `Declare Vector` command, which prompts the user for the size and elements of the vector, iii) use the `VECTOR(f(k),k,n)` command to obtain vector elements $f(k)$, $k = 1, 2, \ldots, n$, or iv) use the `ITERATES` command to obtain vector elements obtained from a recursive sequence.

For example, using the method described in item (iii) above, `Author` `VECTOR(k^3,k,4)` and `Simplify` to obtain to $[1,8,27,64]$. Other stepping parameters for the indices can be used in the `VECTOR` command. See the *Derive User Manual* for details about the parameters of this command.

Matrices are entered in a similar fashion, with vectors acting as rows of the matrix. `Author`

`[[1,2,3],[a,b,c],[x,sinx,6!]]` ,

which results in the matrix

$$1: \quad \begin{bmatrix} 1 & 2 & 3 \\ a & b & c \\ x & \text{SIN} (x) & 6! \end{bmatrix}$$

Similarly, the `Declare Matrix` command prompts for the dimensions and elements of a matrix. Nested calls to the `VECTOR` command can produce a matrix with function evaluations for values of the elements. Additionally, the identity matrix of n dimensions is produced with the `IDENTITY_MATRIX(n)` command.

Descriptions of the basic operators that perform matrix and vector multiplication, transposition, and inversion are provided in the following table:

Operation	Definition
A.B	The . (period or dot) performs vector dot product or matrix multiplication. Vectors must be the same size. The number of columns of A must equal the number of rows of B.
A`	Produces the transpose of matrix A (` is the back accent key, not the apostrophe key).
A^-1	Computes the matrix inverse of matrix A. (A^{-1}).

Matrix and vector multiplication, transposition, and inversion commands.

Descriptions of the most common matrix operations are provided in the following table:

Command	Function
ELEMENT(m,i,j)	Extracts element (i, j) from a vector or matrix m.
CROSS(v,w)	Calculates the cross product of 2 vectors $(v \times w)$.
DIMENSION(m)	Determines the number of rows in matrix m.
OUTER(v,w)	Computes the outer product of vectors v and w.
DET(m)	Calculates the value of the determinant of matrix m.
TRACE(m)	Sums the diagonal elements of matrix m.
ROW_REDUCE(m)	Computes the row echelon form of matrix m.
CHARPOLY(m)	Produces the characteristic polynomial of matrix m.
EIGENVALUES(m)	Determines the eigenvalues of matrix m by finding the roots of the characteristic polynomial.

Matrix manipulation commands and their functions.

To solve a system of linear equations, the equations are entered as elements of a vector and the soLve command is issued. If there are more variables than equations, the system prompts the user for the variables to solve for in the output. If the system is singular, the solution either will contain arbitrary values (@1, @2, etc.) or will display the message No solutions found, depending on whether or not the system is consistent.

Examples of input expressions and results for three linear systems are provided. The input for the first example is to select the Author command and enter

$$\boxed{[\text{x+y-0.4z=10,y-15z=-0.51,-2x-y+z=1}]}\ .$$

The solution is found by executing the menu command soLve. The resulting solution is shown in expression #2 of the following figure.

The second example (expressions #3 and #4) is similar to the first, except expression #3 contains functions instead of equations. In this case, Derive automatically sets the functions equal to 0 to make equations. The solution is shown in expression #4 of the display screen; the output contains @1, which signifies an arbitrary constant.

The third example (expressions #5 and #6) contains four variables in the three equations, so Derive queries the user for the variables to be solved for when the soLve command is issued. In this case a, b, and c are the solve variables given. The solution is given in expression #6 in terms of the variable d.

1:   $[x + y - 0.4\ z = 10,\ y - 15\ z = -0.51,\ -\ 2\ x - y + z = 1]$

2:   $\left[x = -\dfrac{77147}{7600},\ y = \dfrac{157449}{7600},\ z = \dfrac{2151}{1520} \right]$

3:   $[6\ x + 3\ y - z,\ -\ 2\ x - 11\ y + z,\ 7\ x - 14\ y]$

4:   $\left[x = @1,\ y = \dfrac{@1}{2},\ z = \dfrac{15\ @1}{2} \right]$

5:   $[a + b + c = 3\ d,\ 16\ b - 2\ d = 15,\ a - b - 3\ c = 1]$

6:   $\left[a = \dfrac{7\ (10\ d - 1)}{32},\ b = \dfrac{2\ d + 15}{16},\ c = \dfrac{22\ d - 23}{32} \right]$

The ROW_REDUCE command can also be used to solve systems of linear equations. This is a very powerful computational tool in matrix algebra. The operation is performed upon execution of the Simplify command. The ROW_REDUCE command and its computed output to solve the first example are shown.

$$
7: \quad \text{ROW\_REDUCE} \begin{bmatrix} 1 & 1 & -0.4 & 10 \\ 0 & 1 & -15 & -0.51 \\ -2 & -1 & 1 & 1 \end{bmatrix}
$$

$$
8: \quad \begin{bmatrix} 1 & 0 & 0 & -\dfrac{77147}{7600} \\ 0 & 1 & 0 & \dfrac{157449}{7600} \\ 0 & 0 & 1 & \dfrac{2151}{1520} \end{bmatrix}
$$

The utility file VECTOR.MTH contains commands for matrix manipulation. Some of the most powerful commands are presented in the following table:

Command	Function
PIVOT(A,i,j)	Performs one pivoting step in row reducing matrix A.
COFACTOR(A,i,j)	Finds the cofactor of element (i, j) of the matrix A.
EXACT_EIGENVECTOR(A,w)	Finds the eigenvector of matrix A for the exact eigenvalue w. Best used for matrices up to 3×3 in size.
APPROX_EIGENVECTOR(A,w)	Finds the eigenvector of matrix A for the approximate eigenvalue w. Best used for 4×4 and larger matrices.
JACOBIAN(u,θ)	Finds the Jacobian matrix of the system of equations $x = u(\theta_1, \theta_2, \ldots)$.

Some of the matrix manipulation commands in the utility file
VECTOR.MTH and their functions.

Other commands in this file compute outer products; append vectors; and delete, swap, scale, and subtract vector and matrix elements. See the *Derive User Manual* for descriptions of these commands.

1.9   Complex Variables

Complex Variables is a subject which has something for all mathematicians.

—John B. Conway [1973]

Derive can perform complex arithmetic and can handle complex variables and complex-valued functions. Variables are declared complex using the `Declare Variable Domain` command. The imaginary unit i is entered using `#i` or **Alt-i** and displayed as $\hat{\imath}$. Some of the common functions involving complex numbers are given in the following table:

Command	Function
ABS(z)	$|x + \hat{\imath}y|$
SIGN(z)	Finds the point of unit magnitude with the same phase angle as z.
RE(z)	Computes the real part of z, (x).
IM(z)	Computes the imaginary part of z, (y).
CONJ(z)	Finds the complex conjugate of z.
PHASE(z)	Computes the principal phase angle of z.

Common functions involving complex variables.

1.10   Plotting

All the pictures that science now draws of nature ... are mathematical pictures.

—Sir James Hopwood Jeans [1930]

Derive has the capability to do both 2- and 3-dimensional plotting. Both of these capabilities are available through the `Plot` command.

Before the plotting commands are discussed, another important command must be reviewed. Derive has several display modes, so it is important to set the correct mode for your hardware and computational task. The command used to set modes is `Options Display`. If your computer and screen have graphics modes, set the `Mode` to `Graphics` when plotting. The **F5** key switches to the previous display mode, so it is handy to use to switch back and forth between text and graphics modes.

The 2-dimensional `Plot` screen and menu are as follows:

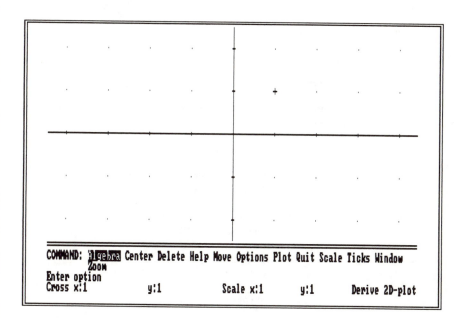

COMMAND: Algebra Center Delete Help Move Options Plot Quit Scale Ticks Window
          Zoom
Enter option
Cross x:1   y:1   Scale x:1   y:1   Derive 2D-plot

To produce a 2-dimensional plot in rectangular coordinates, the following procedure is used: i) highlight the expression to plot (it must be in the form $y = f(x)$ or just $f(x)$, but the variable names do not matter), ii) call the Plot menu, iii) select the location of the plot (Beside, Under, or Overlay), and iv) execute the Plot command.

If three or more functions need to be plotted on the same axes, they can be assembled as components of a vector function. Then all the functions in the vector will be plotted on the same axis at one time.

This produces a plot using the current plotting parameters. The plotting parameters are controlled with additional commands in the Plot menu. The following table gives a short explanation of these Plot commands:

Command	Function
Algebra	Changes control back to the algebra screen.
Center	Positions the center of the plot at the location of the movable cross.
Delete	Deletes functions from the plot list.
Move	Moves the cross to the specified coordinates.
Options	Sets parameters for Accuracy, Color, Display, and Type.
Plot	Plots the functions in the plot list and the highlighted function.
Scale	Sets plotting axes scales to exact values.
Ticks	Establishes the distances between tick marks on the axes. This controls the aspect ratio of the plot.
Window	Opens a new window. This is the same as the Window command in the Algebra window.
Zoom	Automatically changes the plot scale by a fixed amount. Zoom is made In or Out and in the x-, y-, or both directions.
Quit	Stops execution of the Derive program.

2-dimensional Plot menu commands and their functions.

Example: The following example shows a plot of the function $y = f(x)$ defined by

$$y = x + 2\sin x .$$

This function is entered with the **Author** command and by typing

$$\boxed{\text{x+2sinx}} .$$

The **Plot** menu is selected to open the 2-D plotting window. Then the default plot parameter **Scale** is changed to 4 in both the x- and y-directions. Finally, the **Plot** command is issued to produce the following plot:

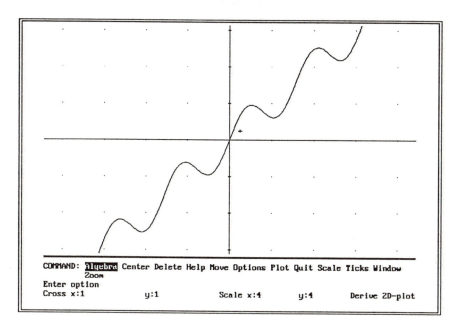

There are several helpful keystrokes and tricks to know while plotting. The direction arrow keys move the small cross around the plot screen. The coordinates of the cross are displayed in the status line at the bottom of the screen. This cross can be handy in finding approximations to many values of the plotted function. The **F9** and **F10** keys are equivalent to the **Zoom In** and **Zoom Out** commands, respectively. Derive will plot three or more functions at the same time when they are entered as components of a vector. Otherwise, you must do the plots one at a time by returning to the algebra window and separately highlighting each function.

Derive also does 2-dimensional polar plotting by changing the coordinate mode to **Polar** using the **Options State** command. In polar mode, the expression is considered to be in the $r = f(\theta)$ or $f(\theta)$ form even though any variable names can be used.

Similarly, 2-dimensional parametric plotting is available by highlighting an expression in the form $[f(t), g(t)]$ with the x (horizontal) direction first and the y (vertical) direction second.

Derive's 3-dimensional rectangular surface plots are produced in a similar manner. The highlighted expression must be in the form $z = f(x, y)$ or just $f(x, y)$. Once again, the actual variable names are not important. When the Plot menu is selected, Derive recognizes the number of variables in the highlighted function and opens the 3-dimensional plot screen. The 3-dimensional plot menu is as shown in the following figure:

```
COMMAND: Algebra Center Eye Focal Grids Hide Length Options Plot Quit Window
         Zoom
Enter option
Center x:0          y:0              Length x:10    y:10      Derive 3D-plot
```

The commands to control the plot parameters for plotting surfaces in three dimensions are described in the following table:

Command	Function
Algebra	Sends control back to the algebra screen and menu.
Center	Positions the center of the plot at the specified coordinates.
Eye	Sets the coordinates of the viewer's eye.
Focal	Sets the coordinates of the focal point.
Grids	Establishes the number of grid panels in the x- and y-directions.
Hide	Allows for removal or inclusion of hidden lines.
Length	Establishes lengths of sides of the box where the plot resides.
Options	Allows control over axes display, colors of lines, and display mode.
Window	Opens a window. This is the same as the Window command in the main menu.
Plot	Produces a plot of the highlighted function.
Zoom	Automatically changes the plot scale by a fixed amount. Zoom can be made In or Out.
Quit	Stops execution of the Derive program.

3-dimensional Plot menu commands and their functions.

The coordinates of the plot's center and length of sides are displayed in the status line at the bottom of the screen.

Example: The following example shows a surface plot of the function $z(x, y)$ defined by

$$z(x, y) = \begin{cases} 5 - |x| - |y| & \text{if } 5 - |x| - |y| > 0 \\ 0 & \text{otherwise} \end{cases}.$$

This function can be converted to a single line expression in several ways using Derive functions. See the descriptions of the MAX and IF commands in Sections 1.6 and 1.14, respectively. Another way to define this function is with the STEP command in Derive. It is entered with the Author command and by typing

$$(5-abs(x)-abs(y)) \ STEP(5-abs(x)-abs(y))$$.

The Plot menu and Overlay location are selected to open the 3-D plotting window. Then the default plot parameter Grid is changed to include 20 grid panels in both the x- and y-directions; the Length parameters are set to $x = 12$, $y = 12$, and $z = 5$; and the Eye parameters are set to $x = 20$, $y = 18$, and $z = 12$. Finally, the Plot command is issued to produce the following surface plot:

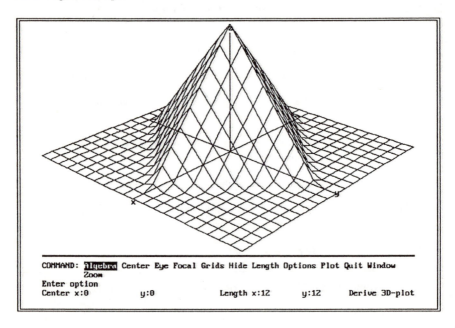

The utility file GRAPHICS.MTH contains functions that plot space curves and parametric surfaces. The actual plotting is done in the 2-D plot window. The AXES command builds the three-dimensional axes in the 2-D plot window. Other commands in the file are ISOMETRIC, ISOMETRICS, SPHERE, TORUS, CYLINDER, CONE, RAYS, and ARCS. *Authoring* and *Simplifying* expressions using these commands produce vectors that can be plotted in the 2-D rectangular plot mode with continuous lines. See the *Derive User Manual* for the specific functions and formats for these commands.

Example: The space curve defined by

$$\vec{r}(t) = \sqrt{t}\cos(4t)\vec{i} + \sqrt{t}\sin(4t)\vec{j} + t/8\vec{k}, \quad 0 < t < 2\pi ,$$

is entered as

ISOMETRIC([sqrt(t)cos(4t),sqrt(t)sin(4t),t/8]) .

The plot of this curve is as follows:

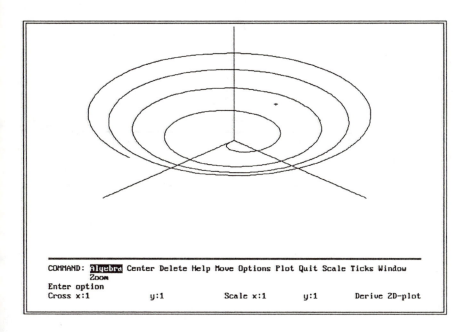

```
COMMAND: Algebra Center Delete Help Move Options Plot Quit Scale Ticks Window
         Zoom
Enter option
Cross x:1            y:1              Scale x:1        y:1         Derive 2D-plot
```

Example: An example plot of a surface in the shape of a cone at an angle of $\pi/8$ radians from the z-axis, centered at the origin, and entered as

ISOMETRICS(CONE(pi/8,t,z),z,-2,2,4,t,-pi,pi,12) ,

is as follows:

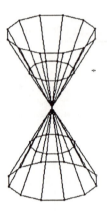

```
COMMAND: Algebra Center Delete Help Move Options Plot Quit Scale Ticks Window
          Zoom
Enter option
Cross x:1              y:1              Scale x:1        y:1        Derive 2D-plot
```

1.11   Numerical Approximations

We propose to use it as a "scientific exploration tool."

—John von Neumann [1946]

When Derive is unable to perform an exact computation or determine a solution symbolically, numerical approximations are viable alternatives. Derive has numerical capabilities in several areas. Some of these capabilities are discussed in this section.

If an antiderivative cannot be determined symbolically for the purpose of evaluating a definite integral, the user can change the Options Precision mode to Approximate to allow Derive to perform numerical quadrature. The numerical method is used automatically when Derive is in Mixed mode and an exact value cannot be calculated. The approximation method used is adaptive Simpson's quadrature. The method approximates the value of the definite integral with an error goal of the digital precision established with the Options Precision command. For example, Derive does not provide an answer to $\int_0^{0.2} e^x/(1 - x^2)\, dx$ in Exact mode. However, in Mixed mode with 6 digits of precision established, the following approximation is produced for this definite integral by using the Author command and entering

INT(Alt-e^x/(1-x^2),x,0,0.2) ,

then executing Simplify.

$$\int_0^{0.2} \frac{\hat{e}^x}{1 - x^2}\, dx$$

$$0.224581$$

A similar circumstance can occur when solving for roots of an expression of one variable. When the equation is too difficult to solve completely in Exact mode, an implicit result is given. If the mode is changed to Mixed or Approximate, Derive uses the bisection method to find a root of the function to within the digital precision established. The user is prompted for the upper and lower bounds of the interval in which to search for the root. For example, if the roots of $e^x - x^4$ are required, in **Exact** mode no solution is produced. However, in either **Mixed** or **Approximate** mode the following result is obtained, if the bounds of 0 and 10 are provided:

$$17: \quad \hat{e}^x - x^4$$

$$18: \quad x = 8.61314$$

The **Calculus Taylor** command finds a Taylor polynomial approximation to an expression. This command can be executed through the menu or in-line as **TAYLOR(f(x),x,x0,n)**, where $f(x)$ is expanded in an nth degree polynomial about x_0. Taylor approximations can be handy for integration when direct symbolic integration is not possible.

It is usually best to use the **Exact** precision mode. When decimal representation of a rational or irrational number is needed, it can be obtained by using the **approX** command.

The utility file SOLVE.MTH has two functions to approximate roots of systems of equations. The function **NEWTONS(u,x,x0,n)** iterates Newton's method n times to find the roots of the vector function $u(x)$ set equal to the zero vector. The vector x_0 contains the initial guess of the roots. The output is a matrix with $n + 1$ rows, where each row is a successive iterate of the method. The **approX** command is used to execute this command.

The other approximation function in this file is fixed point iteration. It is defined by **FIXED_POINT(g,x,x0,n)**. This command produces n iterates of the vector equation $x = g(x)$ with the initial guess x_0. Like the **NEWTONS** function, the resulting output is a matrix with $n + 1$ rows of the successive iterates.

Derive is not designed to perform manipulation of large data sets. However, it has several functions that perform data analysis when the data is placed into a matrix with two or more columns. One such function is **FIT(m)**, which provides a least-squares approximation to the data in the functional form provided. See the *Derive User Manual* for a description of this command.

Another utility file that helps perform numerical operations is NUMERIC.MTH. This utility file contains Derive functions to approximate first and second derivatives of functions with centered difference quotients and to approximate derivatives and integrals of given data arrays. The **INT_DATA(A)** command uses the trapezoidal rule to approximate the antiderivative of data arranged in the matrix **A**. Matrix **A** must contain 2 columns, and each row of **A** represents an $[x, y]$ data point. See the *Derive User Manual* for descriptions of the other commands in this file.

1.12   Utility Files for Calculus

As is well known, physics became a science only after the invention of differential calculus.

—Bernhard Riemann [1882]

In this section, we will discuss the Derive functions found in three utility files: DIF_APPS.MTH, INT_APPS.MTH, MISC.MTH. These functions give Derive added capabilities in performing calculus operations. The help file contains brief descriptions of the functions. Most of the functions in these three files and their purposes are given in the following tables. More information about these commands may be found in the *Derive User Manual*. Many of these functions are used in the examples in Chapter 2.

Function Name	Purpose
CURVATURE(y,x)	Finds the curvature formula for the equation $y = f(x)$.
CENTER_OF_CURVATURE(y,x)	Finds the vector formula for the location of the center of curvature.
TANGENT(y,x,x0)	Finds the equation of the line tangent to $y = f(x)$ at $x = x_0$.
PERPENDICULAR(y,x,x0)	Finds the equation of the line perpendicular to $y = f(x)$ at $x = x_0$.
OSCULATING_CIRCLE(y,x,t)	Finds the parametric representation of the circle that osculates $y = f(x)$ at x.

Basic functions for applying the derivative to find properties of expressions in Cartesian coordinate form found in the file DIF_APPS.MTH.

Function Name	Purpose
PARA_DIF(v,t,n)	Finds $d^n y/dx^n$ in terms of the parametric variable t. v is a vector expression $[x(t), y(t)]$.
PARA_CURVATURE(v,t)	Finds the formula for the curvature of v in terms of t.
PARA_CENTER_OF_CURVATURE(v,t)	Finds the vector expression for the location of the center of curvature in terms of t.
PARA_TANGENT(v,t,t0,x)	Finds the equation in x of the line tangent to the curve $v(t)$ at $t = t_0$.
PARA_PERPENDICULAR(v,t,t0,x)	Finds the equation in x of the line perpendicular to the curve $v(t)$ at $t = t_0$.
PARA_OSCULATING_CIRCLE (v,t,t0,θ)	Finds the parametric representation of the circle that osculates $[x, y] = v$ at t_0.

Basic functions for applying the derivative to find properties of expressions in parametric form found in the file DIF_APPS.MTH.

Function Name	Purpose
POLAR_DIF(r,θ,n)	Finds $d^n y/dx^n$ in terms of the polar variable θ. r is a function of θ.
POLAR_CURVATURE(r,θ)	Finds the formula for the curvature.
POLAR_CENTER_OF_CURVATURE (r,θ)	Finds a vector expression for the location of the center of curvature.
POLAR_TANGENT(r,θ,θ0,x)	Finds the equation in x of the line tangent to $r(\theta)$ at $\theta = \theta_0$.
POLAR_PERPENDICULAR (r,θ,θ0,x)	Finds the equation in x of the (r, θ, θ_0, x) line perpendicular to $r(\theta)$ at $\theta = \theta_0$.
POLAR_OSCULATING_CIRCLE (r,θ,θ0,ϕ)	Finds the parametric representation of the circle that osculates the polar curve r at $\theta = \theta_0$.

Functions for applying the derivative in polar coordinates from DIF_APPS.

DIF_APPS contains two Derive functions that find special properties of 3-dimensional surfaces defined implicitly by $u(x, y, z) = 0$. The following table explains the purpose and form of these two functions that are applications of the derivative:

Function Name	Purpose
TANGENT_PLANE(u,v,v0)	Finds the equation of the plane tangent to $u = 0$ at $v = v_0$. $v = [x, y, z]$.
NORMAL_LINE(u,v,v0,t)	Finds the parametric form of the vector that defines the line perpendicular to $u = 0$ at $v = v_0$.

Functions for finding two properties of a surface from
utility file DIF_APPS.MTH.

DIF_APPS also contains several analogous commands to those commands in the previous three tables for implicit functions defined in Cartesian coordinates $(u(x, y) = 0)$. No table or descriptions are given here for these commands. See the *Derive User Manual* for their descriptions.

The Derive functions that apply or use the integral are located in utility file INT_APPS.MTH. These commands are described in the following table. When a definite integral cannot be evaluated exactly using the Simplify command, you can try a numerical approximation method by selecting the approX command.

Function Name	Purpose
ARC_LENGTH(y,x,x1,x2,u)	Finds the arc length of a curve $y(x)$ with x varying from x_1 to x_2.
POLAR_ARC_LENGTH(r,θ,θ1,θ2,u)	Finds the arc length of a polar curve $r(\theta)$ with θ varying from θ_1 to θ_2.
PARA_ARC_LENGTH(v,t,t1,t2,u)	Finds the arc length of a parametric curve $v(t)$ with t varying from t_1 to t_2.
AREA(x,x1,x2,y,y1,y2,u)	Finds the area between functions $y = y_1(x)$ and $y = y_2(x)$ with x varying from x_1 to x_2.
AREA_CENTROID (x,x1,x2,y,y1,y2,u)	Computes the centroid of the region described above.
POLAR_AREA(r,r1,r2,θ,θ1,θ2,u)	Finds the area between the polar functions $r = r_1(\theta)$ and $r = r_2(\theta)$ with $\theta_1 \leq \theta \leq \theta_2$.
SURFACE_AREA (z,x,x1,x2,y,y1,y2,u)	Finds the surface area or surface integral defined by $z = f(x,y)$.
VOLUME (x,x1,x2,y,y1,y2,z,z1,z2,u)	Finds the volume or volume integral (mass) of a region.
VOLUME_CENTROID (x,x1,x2,y,y1,y2,z,z1,z2,u)	Finds the centroid of a body with density u.
SPHERICAL_VOLUME (r,r1,r2,θ,θ1,θ2,ϕ,ϕ1,ϕ2,u)	Finds the volume integral of a shape defined in spherical coordinates.
CYLINDRICAL_VOLUME (r,r1,r2,θ,θ1,θ2,z,z1,z2,u)	Finds the volume integral of a shape defined in cylindrical coordinates.

Functions for applying the integral found in the
utility file INT_APPS.MTH.

INT_APPS also contains functions for computing moments of inertia and for the advanced techniques of Laplace transform and Fourier series. See the *Derive User Manual* for information about these two functions.

The calculus functions in MISC.MTH are discussed in the following table. This file also contains Derive functions that assist with performing mathematical (inductive) proofs. See the *Derive User Manual* for descriptions of those functions.

Function Name	Purpose
RATIO_TEST(t,n)	Computes a value u. If $u > 1$, $\sum_{n=a}^{\infty} t(n)$ diverges. If $u < 1$, the sum converges. $t(n)$ must be positive.
LIM2(u,x,y,x0,y0)	Evaluates the 2-dimensional limit of $u(x, y)$ as $[x, y] \to [x_0, y_0]$ along a line of slope @1. If the result does not contain @1, the limit is the same for all slopes.
LEFT_RIEMANN(u,x,a,b,n)	Computes the left Riemann sum of $u(x)$ from $x = a$ to $x = b$ using n equally spaced intervals.
INT_PARTS(u,dv,x)	Evaluates the integration by parts formula $\int uv\, dx = u \int(dv)\, dx - \int \left(\int(dv)\, dx \right) \frac{du}{dx}\, dx$.
INT_SUB(y,x,u)	Performs the integration of y by u-substitution.
INVERSE(u,x)	Finds $u^{-1}(x)$.

Functions found in file MISC.MTH.

1.13   Utility Files for Differential Equations

A mathematician, like a painter or poet, is a maker of patterns. If his patterns are more permanent than theirs, it is because they are made with ideas.

—G.H. Hardy [1940]

There are three utility files that contain functions to help solve differential equations. File ODE1.MTH is for first-order equations, and ODE2.MTH contains the functions for second-order equations. The file ODE_APPR.MTH contains numerical approximation methods for first-order differential equations and systems of first-order differential equations. Many of the functions in these utility files are used in the examples in the remaining chapters. Several new functions have been added to these files in Version 2.5 of Derive. If you are using an earlier version of Derive or the Student Edition, refer to section 1.15 of this manual, your *Derive User Manual*, or the book *Derive Laboratory Manual for Differential Equations*[17].

The differential equation to be solved must be placed in proper form in order to identify the arguments for use in the Derive functions that determine its solutions. These built-in functions make Derive an especially powerful and efficient tool for solving many differential equations. Some of the functions in these three utility files and explanations of their purposes are given in the following tables. Further details are provided for many of the functions when they are used in examples in chapters 2 through 7.

---

[17] *Derive Laboratory Manual for Differential Equations* uses Version 1.6 of Derive. It was written by David C. Arney and published by Addison-Wesley. For more references, see the *Other Reading* section near the end of this manual.

Function Name	Purpose
DSOLVE1_GEN(p,q,x,y,c)	Solves differential equation in the form $p(x,y) + q(x,y)y' = 0$.
DSOLVE1(p,q,x,y,x0,y0)	Solves $p(x,y) + q(x,y)y' = 0$. Solution satisfies initial condition $y(x_0) = y_0$.
SEPARABLE_GEN (p,q,x,y,c)	Solves a separable differential equation in the form $y' = p(x)q(y)$.
SEPARABLE(p,q,x,y,x0,y0)	Solves a differential equation in the form $y' = p(x)q(y)$ with $y(x_0) = y_0$.
EXACT_GEN(p,q,x,y,c)	Solves $p(x,y) + q(x,y)y' = 0$, if it is exact. If not exact, returns "inapplicable."
EXACT(p,q,x,y,x0,y0)	Solves an exact equation (above), with given initial conditions $y(x_0) = y_0$.
LINEAR1_GEN(p,q,x,y,c)	Solves $y' + p(x)y = q(x)$.
LINEAR1(p,q,x,y,x0,y0)	Solves a linear equation of the form $y' + p(x)y = q(x)$ with $y(x_0) = y_0$.
HOMOGENEOUS_GEN (r,x,y,c)	Solves $y' = r(x,y)$, if r is homogeneous. If not homogeneous, returns "inapplicable."
HOMOGENEOUS (r,x,y,x0,y0)	Solves $y' = r(x,y)$, if r is homogeneous, with initial conditions $y(x_0) = y_0$.
BERNOULLI_GEN (p,q,k,x,y,c)	Solves Bernoulli equation $y' + p(x)y = q(x)y^k$, where k is a constant.
BERNOULLI (p,q,k,x,y,x0,y0)	Solves Bernoulli equation $y' + p(x)y = q(x)y^k$, with initial condition $y(x_0) = y_0$.
INTEGRATING_FACTOR (p,q,x,y,x0,y0)	Solves an equation when equation times an integrating factor is exact.

Basic functions for solving first-order equations found in ODE1.

Function Name	Purpose and/or Form of Equation to be Solved
DSOLVE2(p,q,r,x,c1,c2)	Solves $y'' + p(x)y' + q(x)y = r(x)$.
DSOLVE2_BV (p,q,r,x,x0,y0,x2,y2)	Solves $y'' + p(x)y' + q(x)y = r(x)$, with boundary conditions $y(x_0) = y_0$ and $y(x_2) = y_2$.
DSOLVE2_IV (p,q,r,x,x0,y0,v0)	Solves $y'' + p(x)y' + q(x)y = r(x)$, with initial conditions $y(x_0) = y_0$ and $y'(x_0) = v_0$.
LIOUVILLE(p,q,x,y,c1,c2)	Solves the Liouville equation $y'' + p(x)y' + q(x)(y')^2 = 0$.
AUTONOMOUS_CONSERVATIVE (q,x,y,x0,y0,v0)	Solves $y'' = q(y)$, with initial conditions $y(x_0) = y_0$ and $y'(x_0) = v_0$.

Functions and the form of applicable second-order equations that can be solved using some methods found in ODE2.

Utility files ODE1 and ODE2 contain several more functions to solve some equations with special forms and more-complicated equations. The following table gives information about the approximation methods found in file ODE_APPR.MTH:

Function Name	Purpose and/or Form of Equation to be Solved
TAY_ODE1(r,x,y,x0,y0,n)	Finds the nth degree Taylor-series solution to $y' = r(x,y)$, $y(x_0) = y_0$.
PICARD(r,p,x,y,x0,y0)	Given an approximate solution p to $y' = r(x,y)$, $y(x_0) = y_0$, finds an improved iterate.
EULER(r,x,y,x0,y0,h,n)	Uses Euler's method to approximate the solution to $y' = r(x,y)$, $y(x_0) = y_0$, at n values of x, $(x_0, x_0 + h, \ldots, x_0 + nh)$. h is the step size of the calculations.
TAY_ODES(r,x,y,x0,y0,n)	Finds the nth degree Taylor-series solution to a system of differential equations.
RK(r,v,v0,h,n)	Uses the Runge-Kutta method to approximate the solution to a first-order equation ($y' = r(x,y)$) or system of such equations for n steps with a step size of h.
DIRECTION_FIELD (r,x,x0,xm,m,y,y0,yn,n)	Produces a matrix of vectors that plots a direction field for $y' = r(x,y)$. There are m vectors in the x-direction, and n vectors in the y-direction.

Functions that solve first-order equations in utility file ODE_APPR.MTH.

In order to use any of these functions, the appropriate utility file (ODE1, ODE2, or ODE_APPR) must be loaded into the work area using the commands **Transfer Load Utility**. Users can make their own utility files for solving differential equations by keeping the most useful commands from these files or new functions they write in the work area and saving them for use in the future using the **Transfer Save** command.

1.14   Programming

We build systems [programs] like the Wright brothers built airplanes—build the whole thing, push it off a cliff, let it crash, and start over again.

—R.M. Graham [1970]

Derive is not a programming language. However, Derive has several commands that provide programming capabilities such as iteration and branching. These formats and purposes of these commands are presented in the following table:

Command	Function
ITERATES(u,x,x0,n)	Computes the iterates of the recursive expression $x_{n+1} = u(x_n)$ for n steps starting with x_0. If the n is omitted, iteration continues until $x_{n+1} = x_n$. The approX command is used to execute this command.
ITERATE(u,x,x0,n)	Performs the same iteration as ITERATES, but returns only the las⁺ value of x_n.
IF(test,t,f,u)	*test* is a conditional test clause that is evaluated as true, false, or unknown. If *test* is true, the *t* expression is evaluated. If *test* is false, the *f* expression is evaluated. If *test* cannot be determined (is unknown), then *u* is evaluated.

Commands that provide Derive with programming capabilities.

The IF command becomes more powerful through the use of nested IF expressions and of logical operators AND, OR, and NOT.

Simple examples of the use of these three commands are presented. If 10 iterates of $x_{n+1} = 2x_n + 6$, $x_0 = -2$, are needed, then Author

$$\boxed{\texttt{ITERATES(2x+6,x,-2,10)}} \; .$$

Simplify to obtain:

```
20:   ITERATES (2 x + 6, x, -2, 10)
21:   [-2, 2, 10, 26, 58, 122, 250, 506, 1018, 2042, 4090]
```

If only the 8th iterate of the above expression is desired, Author

ITERATE(2x+6,x,-2,8) .

Simplify to obtain:

```
22:   ITERATE (2 x + 6, x, -2, 8)
23:   1018
```

Given: the conditional expression in two variables of

$$f(x,y) = \begin{cases} 0 & \text{if } x < 5 \text{ and } y \le 2 \\ x & \text{if } x \ge 5 \text{ and } y > 3 \\ y & \text{otherwise} \end{cases} .$$

To enter this expression, Author

f(x,y):=IF(x<5 AND y<=2,0,IF(x>=5 AND y>3,x,y),y) .

The resulting display is

```
24:  F (x, y) := IF (x < 5 AND y ≤ 2, 0, IF (x ≥ 5 AND y > 3, x, y), y)
```

Now in order to evaluate the expression for various values of x and y, the function f(x,y) is *Authored* and *Simplified*. For example, the input and output for the three pairs of x, y points (4,1), (6,4) and (2,4) is as follows:

```
25:  F (4, 1)

26:  0

27:  F (6, 4)

28:  6

29:  F (2, 4)

30:  4
```

The IF command can help in plotting discontinuous functions. For example, if the function g of one variable x is defined by

$$g(x) = \begin{cases} x & \text{if } x < -1 \\ 2 & \text{if } -1 \leq x \leq 1 \\ -x & \text{if } x > 1 \end{cases},$$

the command to Author the expression is

$$\boxed{\text{IF(x<-1,x,IF(x<=1,2,-x))}} .$$

This function can then be plotted by issuing Plot Plot. The following plot of this function was produced with an accuracy setting of 7:

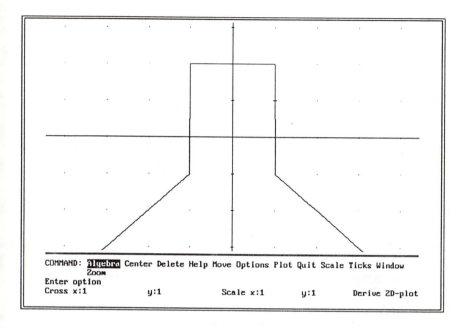

With this accuracy setting, the discontinuities at $x = -1$ and $x = 1$ are not obvious, since the steep line connects the points on both sides of the discontinuity. By changing the accuracy setting to 9 with the Options Accuracy command and replotting, the following plot is produced. This new plot accurately shows the discontinuities of the function.

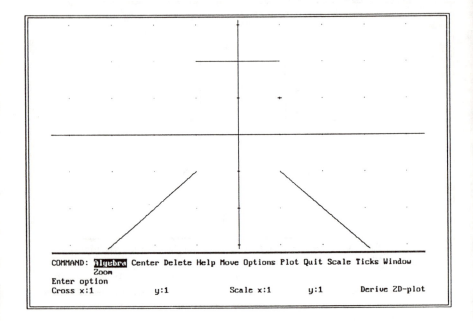

```
COMMAND: Algebra Center Delete Help Move Options Plot Quit Scale Ticks Window
         Zoom
Enter option
Cross x:1          y:1          Scale x:1      y:1        Derive 2D-plot
```

1.15   Other Versions of Derive

*The science of figures, to a certain degree, is not only indispensably
requisite in every walk of civilized life, but the investigation of
mathematical truths accustoms the mind to method and correctness in
reasoning, and is an employment perculiarly worthy of rational beings.*

—George Washington [1788]

This guide is designed for use with Version 2.5 of Derive. There are several
other versions of Derive available. All are very similar and can be used with this
guide by making slight modifications to some of the commands. Version 1.6 or
later issues of Version 1 contain different functions to solve differential equations
in the ODE1, ODE2, and ODE_APPR utility files. The book *Derive Laboratory
Manual for Differential Equations*, written by the author of this book in 1991,
presents similar material to that found in this book for users of Version 1 of Derive.
Version 2 (earlier than 2.5) also has slightly different functions in the utility files to
solve ODEs. The ROM-card version of Derive for the Hewlett-Packard 95LX and
100LX Palmtop computers has both a Version 2 and a Version 2.5, with slightly
different formatting for the smaller screen size. Another version of Derive, Derive
XM was introduced in February of 1993. The instructions in this book will work
with that version as well. Derive XM is identical to the original, menu-driven program, providing the same features and functions for solving symbolic and numeric
problems. In addition Derive XM can make use of up to 4 gigabytes of extended
memory to solve much larger problems. Derive XM requires a 386 or 486 based
PC compatible computer with 2 or more megabytes of extended memory.

If you have an earlier, registered version of Derive, you may be able to update
to the latest version at a reduced cost by contacting your dealer. In any case, this
manual can be used with any version of Derive with only minor adjustments.

1.16   Limitations

What we know is very slight. What we don't know is immense.

—Laplace [1827]

All software packages have limitations, and Derive is no exception. It is usually best to be warned of those limitations before discovering them firsthand at a critical step in solving a problem.

Derive's most obvious restriction is probably its limited programming capability. Derive's commands are executed one at a time with user interface needed after each step. There is no way to execute loops of multi-line commands. However, through the use of nested functional calls, several operations can be executed at once. Despite this limitation, there are ways to do iteration (ITERATES command), recursion, and branching (IF command). See Section 1.14 for a discussion of these commands. An inconvenience related to this situation is that subscripted variables use the ELEMENT function to access elements.

Derive does not support the use of a mouse to facilitate the interface with the menu. This sometimes slows cursor movement, highlighting, and selection of menu options.

Derive's work area is designed for executable commands, not text. Therefore, there is little text formatting provided in the work area. This restricts the quality of comments in the screen image and printed output. Also, long expressions run off the screen and are difficult to read.

The limitations in Derive's plotting tools are the *lack* of i) direct entry of domain bounds for a rectangular plot, ii) an automatic scaling option for two-dimensional plots, iii) the ability to label axes and designate the axes to specific variables, iv) the ability to handle implicit functions, and v) the ability to plot level curves for a surface. Plotting sometimes seems slow in Derive. However, the user has the ability to change accuracy parameters to speed up plotting when precision can be sacrificed for speed. The ability to plot a vector of expressions with one plot command is very helpful and should be used if several plots on the same axes are needed.

1.17   Recommendations

*Old people like to give good advice, as solace for no longer being able to
provide bad examples.*

—François, Duc de la Rochefoucald [1678]

There are several things that can be done to ensure efficient use of the
tremendous capabilities of Derive and to tailor the software to your specific
needs. First, ensure the system state you use most often is saved into the
DERIVE.INI file and, therefore, is in effect when the program is started. This
is accomplished by putting the computer in the state you want and then saving
that state using **Transfer Save State**. If you use the plotting features often,
ensure this initial state is in the appropriate **Graphics** mode. This eliminates
the need to change states in order to get the capabilities you need. See Sections
1.2 and 1.5 and the *Derive User Manual* for more information.

Next, build and save utility files that contain the commands that you use
in a manner that is efficient for your use. For instance, select the commands
you use most often in the utility files DIF_APPS.MTH, INT_APPS.MTH,
ODE1.MTH, ODE2.MTH, and ODE_APPR.MTH and design your own utility
file for solving differential-equation problems. You may want to add other
functions you frequently use from MISC.MTH or SOLVE.MTH to your
personal utility file. You may even want to build several special utility files.

The following table lists functions, along with the Derive utility file they
are found in, that may produce a nice file for a differential-equations course:

Function Name	Utility File Location of Function
DSOLVE1	ODE1
SEPARABLE	ODE1
EXACT	ODE1
LINEAR1	ODE1
DSOLVE2	ODE2
DSOLVE2_BV	ODE2
DSOLVE2_IV	ODE2
EULER	ODE_APPR
RK	ODE_APPR
TAY_ODE1	ODE_APPR
PICARD	ODE_APPR
LAPLACE	INT_APPS
FOURIER	INT_APPS

Suggested functions for a utility file for an elementary
differential-equations course.

If you frequently repeat a sequence of commands or use a command in a special way, try to design your own commands and add them to appropriate utility files.

Finally, spend some time learning the structure of the menus and the keystrokes that help manipulate expressions and reduce retyping expressions. The menu structure is discussed in Section 1.2. The keystrokes are discussed in Section 1.3. Derive's menu system makes it user-friendly; however, the understanding and use of a few of the special keystrokes can turn you into a more efficient user of this powerful software tool.

Probably the most overlooked and underutilized features are i) using **F3**, **F4**, and expression labels (such as #21) in the author line; ii) using **Ctrl-S** and **Ctrl-D** to move the cursor within the author line; iii) highlighting a subexpression (part of an existing expression) before using Simplify, Expand, Factor, approX, or Manage Substitute; and iv) using the approX command instead of Simplify when desiring a numerical approximation.

1.18   Demonstration of Basic Capabilities

Computing offers a tool with which mathematics influences the modern world and a means of putting mathematical ideas into action.

—Lynn Arthur Steen [1990]

This section presents demonstrations of many of Derive's commands, functions, and capabilities. For the most part, the problems solved here are simple drill problems that can be solved directly by a single Derive command. The usual method used for these problems is to Author an expression and then Simplify the expression. The format for each demonstration problem is as follows: 1) problem statement, 2) Derive input, outlined in a box, 3) the output display from Derive that shows the solution of the problem, and 4) information and comments about the problem, when needed.

Demonstration #1. Expand $(x + y)^5$.

Author $\boxed{(x+y)^5}$. Press **Enter**. Select Expand and press **Enter** (to accept the default value) for the **Expand Variable**.

$$1: \quad (x + y)^5$$

$$2: \quad x^5 + 5 x^4 y + 10 x^3 y^2 + 10 x^2 y^3 + 5 x y^4 + y^5$$

Demonstration #2. Factor $x^4 - x^3 - 19x^2 - 11x + 30$.

`Author` ┌─────────────────────────┐ `x^4-x^3-19x^2-11x+30` └─────────────────────────┘ . Press **Enter**. Select `Factor`. Then select `Complex`.

$$3: \quad x^4 - x^3 - 19\,x^2 - 11\,x + 30$$

$$4: \quad (x - 5)\,(x - 1)\,(x + 2)\,(x + 3)$$

Demonstration #3. Find the prime factors of 44566469030160.

`Author` the number ┌──────────────────┐ `44566469030160` └──────────────────┘ . Press **Enter**. Then select the `Factor` command.

$$5: \quad 44566469030160$$

$$6: \quad 2^4 \; 3^5 \; 5 \; 7 \; 11 \; 1621 \; 18367$$

Demonstration #4. Evaluate and approximate $\ln(\sqrt{2})$.

Author $\boxed{\texttt{ln(sqrt(2))}}$. Press **Enter**. Select `Simplify`.

```
    7:    LN (√2)

          LN (2)
    8:    ──────
             2
```

Execute `approX`.

```
    7:    LN (√2)

          LN (2)
    8:    ──────
             2

    9:    0.346573
```

The format of the output depends upon the settings selected in the `Options Notation` submenu. The options set of this output are highlighted on the following menu display:

```
NOTATION: Style: Decimal Mixed Rational Scientific   Digits: 6

Enter numerical output style
Approx(8)                                   Free:96%
```

Demonstration #5. Find $\lim_{x \to 0} \sin(x)/x$.

Author the Derive statement $\boxed{\text{LIM(sinx/x,x,0)}}$. Press **Enter**. Select the Simplify command.

```
                  SIN (x)
      10:   lim  ─────────
            x→0      x

      11:   1
```

Another method to evaluate limits can be performed using the Calculus menu command (see Section 1.7).

Demonstration #6. Find the first derivative of $x \sin(3x^3 - x)$.

Author the function $\boxed{\text{DIF(x sin(3x^3-x),x)}}$. Press **Enter**. Select the Simplify command.

```
            d              3
      12:   ── x SIN (3 x  - x)
            dx

                   2           3                 3
      13:   x (9 x  - 1) COS (3 x  - x) + SIN (3 x  - x)
```

Another method to evaluate derivatives can be performed using the Calculus menu command. See Sections 1.7 and 2.3.

Demonstration #7. Find the second derivative with respect to x of $|x| \cos x - e^{ax}$.

Author $\boxed{\text{DIF(abs(x)cosx-Alt-e\^{}(ax),x,2)}}$. Press **Enter**. Select Simplify.

14: $\left[\dfrac{d}{dx}\right]^2 \; (|x| \; \text{COS} \; (x) - e^{a \, x})$

15: $- \text{SIGN} \; (x) \; (x \; \text{COS} \; (x) + 2 \; \text{SIN} \; (x)) - a^2 \; e^{a \, x}$

Demonstration #8. Find $\int x \cos^2 x \, dx$.

Author $\boxed{\text{INT(xcos\^{}2(x),x)}}$. Press **Enter**. Select Simplify.

16: $\displaystyle\int x \; \text{COS} \; (x)^2 \; dx$

17: $\dfrac{x \; \text{SIN} \; (x) \; \text{COS} \; (x)}{2} - \dfrac{\text{SIN} \; (x)^2}{4} + \dfrac{x^2}{4}$

Another method to evaluate integrals can be performed using the Calculus menu command. See Sections 1.7 and 2.4.

Demonstration #9. Find $\int_1^\infty \frac{1}{x^2}\,dx$.

Author $\boxed{\text{INT(1/x\^2,x,1,inf)}}$. Press **Enter**. Select **Simplify**.

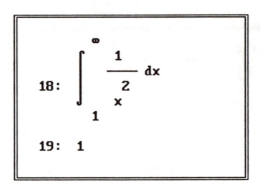

Demonstration #10. Find the 5-term Taylor polynomial approximation to $e^{-x}\cos(x^2)$ about the point $x = 0$.

Author $\boxed{\text{TAYLOR(\textbf{Alt-e}\^-x cos(x\^2),x,0,5)}}$. Press **Enter**. Select **Simplify**.

$$22:\quad \text{TAYLOR}\left(\hat{e}^{-x}\ \text{COS}(x^2),\ x,\ 0,\ 5\right)$$

$$23:\quad \frac{59\,x^5}{120} - \frac{11\,x^4}{24} - \frac{x^3}{6} + \frac{x^2}{2} - x + 1$$

Another method to determine Taylor polynomials can be performed using the `Calculus` menu command. See Sections 1.7 and 4.3.

Demonstration #11. Evaluate the function $\sqrt{x}\,e^{2x}\csc(\pi x)$ for $x = 4.5$.

Author $\boxed{\text{sqrt(x)Alt-e\^{}(2x)csc(pi x)}}$. Press **Enter**. Execute Manage Substitute.

Press **Enter**. Replace x with $\boxed{4.5}$. Select Simplify.

```
                  2 x
       24:  √x ê       CSC (π x)


                  2 4.5
       25:  √4.5 ê       CSC (π 4.5)


               9
           3 √2 ê
       26:  ─────────
              2
```

Demonstration #12. Evaluate $2^{32} - 1$ and factor it into prime factors.

Author $\boxed{\text{2\^{}32-1}}$. Press **Enter**. Select Expand. Select Factor.

```
               32
       27:  2    - 1

       28:  4294967295

       29:  3 5 17 257 65537
```

Demonstration #13. Perform the partial fraction expansion of the rational expression $\frac{x^2-7x-3}{x^3+x^2-8x-12}$.

Author $\boxed{\text{(x^2-7x-3)/(x^3+x^2-8x-12)}}$. Press **Enter**. Select **Expand**.

$$30: \quad \frac{x^2 - 7x - 3}{x^3 + x^2 - 8x - 12}$$

$$31: \quad -\frac{3}{(x+2)^2} + \frac{8}{5(x+2)} + \frac{3}{5(3-x)}$$

2

Basic Tools

The heart of mathematics consists of concrete examples and concrete problems.

—Paul R. Halmos [1970]

In this chapter, several sample problems have been worked out using some of the basic graphing and calculus functions to show the power and versatility of Derive as a problem-solving tool. In carefully reading these problems and solutions and working along with Derive, the reader should get a better feel for the basic operation and capabilities of Derive. Some of these examples also show the limitations of the computer package, and help the reader to know when to use Derive and the computer and when not to use these tools.

The exercises in this chapter are to be done by you, the reader. These problems direct you to learn new things about Derive's basic capabilities. Some of the exercises refer to the examples as a source of help. Some exercises lead you to explore the mathematics and software and to discover new results on your own. As you solve the questions from these exercises, ask yourself "what if" questions and answer them. Good luck, and have fun exploring.

Example 2.1: Curve Sketching

A picture shows me at a glance what it takes dozens of pages of a book to expound.

—Ivan Turgenev [1862]

Subject: Curve Sketching and Analyzing Function Behavior

References: Sections 1.6 and 1.10

Problem: Sketch the following functions and categorize their local and global behavior:

$$f(x) = \frac{1}{x}\sin(x^2),$$

$$g(x) = \sin x + x,$$

$$h(x) = \tanh(1 - 20x),$$

$$s(x) = \text{Floor}(x^2, 1) = \lfloor x^2 \rfloor = \text{integer part of } x^2.$$

Solution: Often the most important step in solving a problem is sketching the curve or function. Derive and Calculus complement each other as aids to curve sketching, especially as it relates to finding the most important features of a curve—local maximums, local minimums, intercepts, and asymptotes. Producing plots in Derive is easy. The challenge is to produce a plot of the function with the correct scale and viewing window to gain good geometrical intuition about the behavior of the function.

Be sure your computer is in the correct display mode. Derive has several possible display modes. The command used to set the modes is `Options Display`. If your computer and monitor have graphics modes, set the `Mode` to `Graphics` when plotting. You may also want to change the colors used in the plotting screen. The `Options Color` command is used to set the colors for the various parts of the screen. See Section 1.10 for more information on plotting modes.

The first step to plot a function is to `Author` the function into the work area. The entry of $f(x)$ is $\boxed{\texttt{1/x sin(x\^2)}}$. Then select the `Plot` submenu. There are three options for the location of the plotting window: `Beside`, `Under`, or `Overlay`. In this case, to use the entire screen for plotting choose `Overlay`. The actual plot region you see in the display depends upon the plotting parameters set with the commands in the `Plot` submenu.

Before we plot $f(x)$, let's reset the plot parameters to try to capture the global behavior of $f(x)$. First, we will center the plotting region at (0,0). Do this by executing Move and setting $x = 0$ and $y = 0$. Remember that the Tab key provides for the movement between parts of the menu. Then select Center to make (0,0) the center of the plotting region. In order to obtain a global perspective by seeing more of the change in the x-direction, the x-axis scale must be increased to a higher value than the y-axis scale. Execute the Scale command and set the x-axis scale to 3 and the y-axis scale to 1. Now select Plot to acquire the following graph:

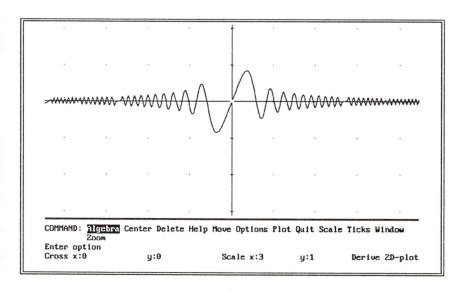

This function seems to oscillate with decaying amplitude the further it gets from the origin. We also notice some strange behavior (oscillations are interrupted slightly) at a couple of places. Further investigation, by zooming in and out (using F9 and F10 keys) at those locations, shows this to be aliasing caused by the graphing procedure and the scale being used. Aliasing is the appearance of a behavior on the graph that is not actually present in the function. It is caused by interpolation in the graphing procedure along with the large scale being used.

To get a plot of the local behavior near a point of interest, we need to change the scale and move the center of the plot. Rescale as we did before. This time set x-scale = 0.25 and y-scale = 0.25. Let's see the behavior near $x = 10$; so select the Move command and set $x = 10$ and $y = 0$. Issue the Center command. The plot is redrawn with these plotting parameters. At this interval and scale, the function locally looks like a normal sine curve with a reduced amplitude.

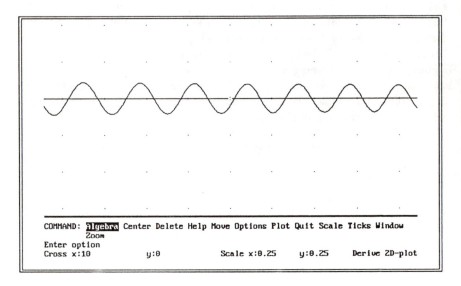

Execute Delete All to clear the plot window. Select Algebra to return control to the algebra window.

Let's attack $g(x)$ in a similar fashion. Author ⟨sinx+x⟩. Issue Plot to obtain the 2-dimensional plotting window. For a global view of the function, select Move to move the cross to $x = 0$ and $y = 0$, and execute Center. Set the scale using the Scale command to $x = 10$ and $y = 10$ per tick mark on the axes. This provides a plotting region of approximately $-40 < x < 40$ and $-30 < y < 30$. Issue the Plot command to obtain the following graph of $y = g(x)$:

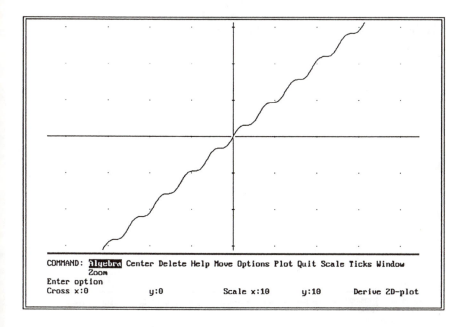

```
COMMAND: Algebra Center Delete Help Move Options Plot Quit Scale Ticks Window
         Zoom
Enter option
Cross x:0              y:0              Scale x:10      y:10       Derive 2D-plot
```

The graph of $g(x)$ globally resembles the straight line $y = x$. Larger scales will eliminate the wiggles altogether.

To investigate the local behavior near $(0,0)$, we must rescale the tick marks. Select the **Scale** submenu and set $x = 1.5$ and $y = 1.5$. The **Tab** key moves control through different parts of the submenu. The plot is automatically redrawn with the new scale when the **Enter** key is pressed.

The plot showing the local behavior is as follows:

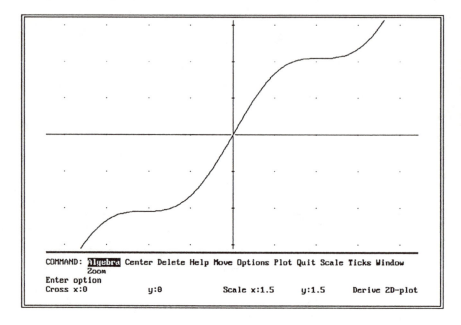

```
COMMAND: Algebra Center Delete Help Move Options Plot Quit Scale Ticks Window
         Zoom
Enter option
Cross x:0              y:0              Scale x:1.5      y:1.5      Derive 2D-plot
```

.Return to the **Algebra** window by selecting the **Algebra** command. In order to close the plotting window, the **Window Close** command is used.

For the function $h(x)$, we will do something a little different. We will split the screen into two windows: one to show the sketch of $h(x)$ and the other to display the functional definition. First **Author** $\boxed{\texttt{tanh(1-20x)}}$. Then select **Window Split Horizontal**. Then answer the query to split the screen at line 5. Notice that the two windows are numbered in the upper-left corners. The top window (#1) has its number highlighted, which indicates that window is the active one. The **F1** key toggles control between the two windows. This kind of window can also be established with the **Plot Under** command, when selecting the plot's location. (See Sections 1.4 and 1.10 for more information about using windows and performing plotting.)

Press **F1** to activate window #2. Now we will designate this window for plotting with the **Window Designate 2D-Plot** command. Answer **Y** for yes to the query to abandon the current algebraic expressions in this window. The algebraic expressions will remain in window #1. The screen is now split, with part of the screen set up for plotting, and appears as follows:

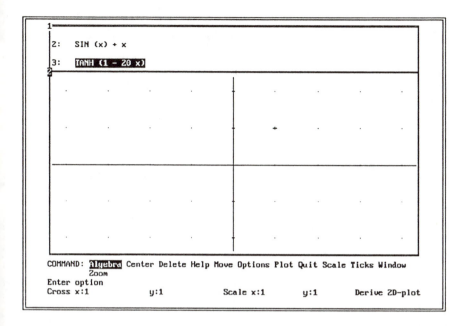

Now we can perform the graphing of $h(x)$ by executing Plot. The resulting graph is shown.

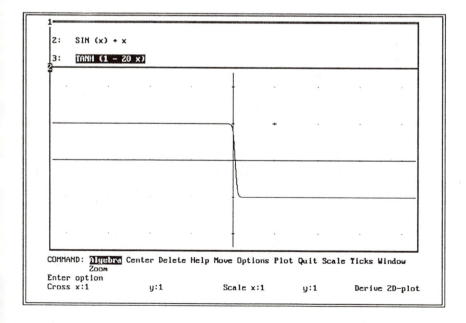

Globally this function resembles a step function. It appears to be 1 for $x < 0$ and -1 for $x > 0$ with a very steep, near-vertical line connecting these values near or at $x = 1$. Before we jump to any conclusions, let's get a closer look near $x = 0$. Select Scale and set the x-scale to 0.05. Press **Enter** to obtain this new, smaller-scaled plot.

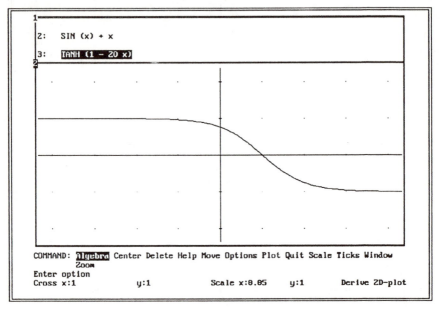

On this finer-scaled plot, we see that the transition from 1 to -1 takes place very smoothly and the x-intercept is not 0, but a value close to 0.05. Actually, $h(x)$ is locally quite different from a discontinuous step function. Close the plot window using the Window Close command.

Now we are ready to tackle $s(x)$. The FLOOR function is in the utility file MISC.MTH, so we must use the Transfer Load Utility commands to load the functions from that utility file into the work area. Enter Misc for the file name and notice Derive loading all the commands available in that file. Then Author

$$\boxed{\text{FLOOR(x\^2,1)}} \;.$$

Then execute Plot and Overlay. Select Scale and set x-scale to 1 and y-scale to 5. The Options Accuracy must be set to 9 to get a pointwise accurate plot.

Then execute Plot to get

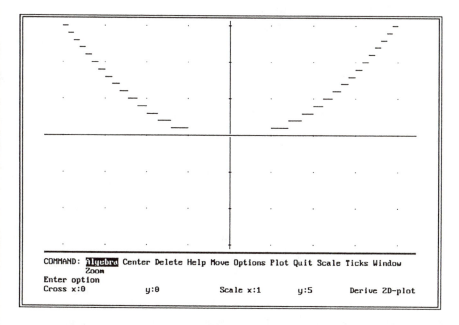

```
COMMAND: Algebra Center Delete Help Move Options Plot Quit Scale Ticks Window
         Zoom
Enter option
Cross x:0              y:0              Scale x:1      y:5        Derive 2D-plot
```

Try other settings for the accuracy parameter to see the problems that can occur in plotting discontinuous functions.

Example 2.2: Root Finding

A great discovery solves a great problem but there is a grain of discovery in the solution of any problem.

—George Polya [1946]

Subject: Finding or Approximating Roots of Functions

References: Sections 1.6, 1.7, and 1.10

Problem: Find all the real zeros of the following functions:

$$f(x) = x^3 + 4x^2 - .6x + 3.7,$$

$$g(x) = x^5 + 7x^4 - 0.2x^3 - x^2 + 0.5x - 10,$$

$$h(x) = e^x - x \sin x - 2.$$

Solution: Finding the roots of some equations or the zeros of some functions can be quite simple. First, enter the equation or function into the work area with the Author command. For the function $f(x)$, the expression is entered as $\boxed{\text{x\^3+4x\^2-0.6x+3.7}}$. Then just execute soLve by typing L, the letter that is shown in uppercase in the command. The result is

As expected, this cubic polynomial has three roots. However, in this case only one is real. The other two are complex, and all three are messy. The two complex roots are truncated in the figure and on the screen. The direction-arrow keys can be used to see the portion of the expression off the right of the screen. Highlight each root in turn and execute approX. The three approximate roots in decimal form are as follows:

```
5:    x = -4.33526

6:    x = 0.167632 - 0.908494 î

7:    x = 0.167632 + 0.908494 î
```

Now we will try the same procedure for the fifth-degree polynomial $g(x)$.
Author

```
x^5+7x^4-0.2x^3-x^2+0.5x-10
```

and execute soLve. This time Derive doesn't help us.

Before turning to an approximate numerical procedure, we may be able
to approximate some of the roots using graphics. Move the highlight to the
expression for $g(x)$ through the use of the direction-arrow keys (↑ and ↓).
Execute the Plot Overlay command. Then select Scale and set x-scale $= 2$
and y-scale $= 5$. Then execute Plot. The resulting plot with the designated
scale parameters is

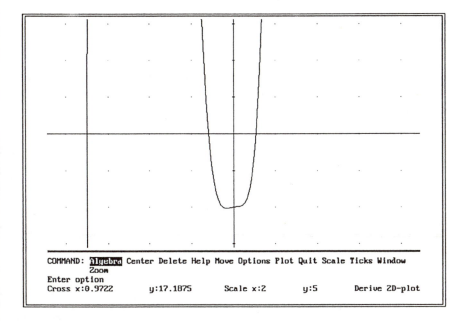

Carefully move the cross to a point near the three x-intercepts, or roots.
We come up with the following three approximate values for the roots: -7,

−1.2, and 1.1. Now change the mode from **Exact** to **Approximate** using the **Options Precision** command. Set the accuracy to 6 digits. Now the **soLve** command performs the bisection approximation procedure. Return to the algebra window and execute **soLve**. To find the root near $x = -7$, provide lower and upper values of −8 and −6, respectively. The result is

```
        5       4         3     2
8:    x   + 7 x   - 0.2 x   - x   + 0.5 x - 10

9:    x = -7.00256
```

Repeat this same procedure to find the other two roots of the polynomial using the bisection search intervals of $(-1.5, -1)$ and $(0.9, 1.1)$. The results are as follows:

```
9:    x = -7.00256

10:   x = -1.19065

11:   x = 1.07710
```

Finding all the real roots of function $h(x)$ is also difficult. **Author** the command $\boxed{\textbf{Alt-e\^x-x-2}}$. Then issue the **Plot Plot** command to obtain the following Derive output:

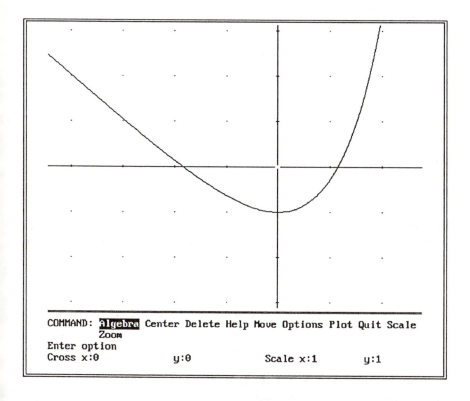

We see that $h(x)$ has one positive and one negative root. Return to the Algebra window. Change the precision to Approximate using the Options Precision command. Then execute soLve. In order to obtain the negative root, input -4 for the lower bound and 0 for the upper bound. To obtain the positive root, execute soLve with a lower bound of 0 and an upper bound of 4. The results of these operations are as follows:

```
12:  ê  - x - 2

13:  x = -1.84140

14:  x = 1.14619
```

Example 2.3: Derivatives

But the velocities of the velocities—the second, third, fourth, and fifth velocities, etc.—exceed, if I mistake not, all human understanding.

—George Berkeley [1734]

Subject: Using Derivatives to Analyze Functions

References: Sections 1.7 and 1.11

Problem: Use the appropriate derivatives and analysis to find the points of local and global maximum and minimum and the points of inflection for the following functions:

$$f(x) = x^3 - 4x^2 - 6x - 1/2,$$
$$g(x) = \frac{x^2 + 1}{x^2 - 4x + 3},$$
$$h(x) = x \ln x + \sin x.$$

Solution: Finding the derivative of functions using Derive is simple and direct. There are two ways to obtain a derivative of a function. See Section 1.7 for more information. The first way is to **Author** the function into the work area. Then select the **Calculus** menu. Select **Differentiate** and input the parameters for order and variable of differentiation. The keystrokes for finding $\frac{df}{dx}$ are as follows: **Author**

$$\boxed{\texttt{x\^{}3-4x\^{}2-6x-1/2}} \ .$$

Execute **Calculus Differentiate** and respond with variable of differentiation x and order of 1. Then execute **Simplify**.

The resulting output is as follows:

$$1:\quad x^3 - 4x^2 - 6x - \frac{1}{2}$$

$$2:\quad \frac{d}{dx}\left[x^3 - 4x^2 - 6x - \frac{1}{2}\right]$$

$$3:\quad 3x^2 - 8x - 6$$

To find some of the critical values, we set $\frac{df}{dx} = 0$ and solve for the roots. We find these roots with Derive by executing the soLve command. The expression is assumed to be set equal to 0 for this operation. The resulting two roots (critical values) are shown in the following display:

$$3:\quad 3x^2 - 8x - 6$$

$$4:\quad x = \frac{4}{3} - \frac{\sqrt{34}}{3}$$

$$5:\quad x = \frac{\sqrt{34}}{3} + \frac{4}{3}$$

The second derivative $\frac{d^2 f}{dx^2}$ can be obtained by taking the derivative of $\frac{df}{dx}$. Simply highlight the expression for $\frac{df}{dx}$ using the arrow keys and issue the Calculus Differentiate command. Simplify performs the differentiation and gives the following result:

$$6:\quad \frac{d}{dx}(3x^2 - 8x - 6)$$

$$7:\quad 6x - 8$$

Now use the Manage Substitute command to find the value of the second derivative at the critical points. These values are rather complicated and long to retype, so use the highlight (direction keys) and the **F3** key to obtain the values for the replacement of x. After the first value is substituted, execute approX to get the evaluation of $\frac{d^2 f}{dx^2}$ at the first critical value. The result is negative, as shown in the following display:

$$
\begin{array}{ll}
7: & 6\,x - 8 \\[2ex]
8: & 6 \left[\dfrac{4}{3} - \dfrac{\sqrt{34}}{3} \right] - 8 \\[3ex]
9: & -11.6619
\end{array}
$$

Therefore, this critical point produces a local maximum. This can be verified graphically using the plotting techniques of Example 2.1.

Now we go back to the expression for the second derivative and Manage Substitute the second critical value into the function. Issue the approX command to get

$$
\begin{array}{ll}
10: & 6 \left[\dfrac{\sqrt{34}}{3} + \dfrac{4}{3} \right] - 8 \\[3ex]
11: & 11.6619
\end{array}
$$

This time the value is positive and we have a local minimum.

To find the inflection points, the roots of $\frac{d^2 f}{dx^2} = 0$ are needed. Highlight the expression for the second derivative and execute soLve. The result is the inflection point:

$$
12: \quad x = \frac{4}{3}
$$

Let's analyze $g(x)$ in a similar manner. However, we will use a different procedure to set up the derivatives. The second method to find derivatives uses the in-line DIF command. See Section 1.7 for the description of this command. To find $\frac{dg}{dx}$, Author

$$\boxed{\texttt{DIF((x\^2+1)/(x\^2-4x+3),x)}}\ .$$

The order defaults to first-order when no third argument is entered. Simplify to get

$$13: \quad \frac{d}{dx} \ \frac{x^2 + 1}{x^2 - 4\,x + 3}$$

$$14: \quad - \ \frac{4\,(x^2 - x - 1)}{(x^2 - 4\,x + 3)^2}$$

We find the roots of $\frac{dg}{dx} = 0$ by executing soLve. When the soLve command is given an expression that is not an equation or an inequality, it returns the zeros of the expression. The result is as follows:

$$14: \quad - \ \frac{4\,(x^2 - x - 1)}{(x^2 - 4\,x + 3)^2}$$

$$15: \quad x = \frac{1}{2} - \frac{\sqrt{5}}{2}$$

$$16: \quad x = \frac{\sqrt{5}}{2} + \frac{1}{2}$$

The second derivative $\frac{d^2g}{dx^2}$ can be determined with a similar procedure. Author

$$\boxed{\texttt{DIF((x\^2+1)/(x\^2-4x+3),x,2)}}\ .$$

The 2 in the third argument of this command indicates second-order. Simplify to obtain the following:

$$17: \quad \left[\frac{d}{dx}\right]^2 \frac{x^2 + 1}{x^2 - 4x + 3}$$

$$18: \quad \frac{4\,(2\,x^3 - 3\,x^2 - 6\,x + 11)}{(x^2 - 4\,x + 3)^3}$$

Manage Substitute the critical values into the second derivative one at a time and execute approX to get the values of the second derivative at the critical values. The following display shows these operations for the two critical values:

$$18: \quad \frac{4\,(2\,x^3 - 3\,x^2 - 6\,x + 11)}{(x^2 - 4\,x + 3)^3}$$

$$19: \quad \frac{4\left[2\left[\frac{1}{2} - \frac{\sqrt{5}}{2}\right]^3 - 3\left[\frac{1}{2} - \frac{\sqrt{5}}{2}\right]^2 - 6\left[\frac{1}{2} - \frac{\sqrt{5}}{2}\right] + 11\right]}{\left[\left[\frac{1}{2} - \frac{\sqrt{5}}{2}\right]^2 - 4\left[\frac{1}{2} - \frac{\sqrt{5}}{2}\right] + 3\right]^3}$$

$$20: \quad 0.260990$$

$$21: \quad \frac{4\left[2\left[\frac{\sqrt{5}}{2} + \frac{1}{2}\right]^3 - 3\left[\frac{\sqrt{5}}{2} + \frac{1}{2}\right]^2 - 6\left[\frac{\sqrt{5}}{2} + \frac{1}{2}\right] + 11\right]}{\left[\left[\frac{\sqrt{5}}{2} + \frac{1}{2}\right]^2 - 4\left[\frac{\sqrt{5}}{2} + \frac{1}{2}\right] + 3\right]^3}$$

$$22: \quad \boxed{-12.2609}$$

We see by the results that $x = 1/2 - \sqrt{5}/2$ is the value where the function attains a local minimum, and $x = 1/2 + \sqrt{5}/2$ is the value where the function attains a local maximum.

In order to find the points of inflection, highlight the expression for $\frac{d^2 g}{dx^2}$ and execute soLve. The result is

```
         1/3       2/3
        5         5        1
23:  x = - ─────── - ───── + ─
           2         2       2


         1/3       2/3                   1/6         1/6
        5         5        1        ┌  16875       675    ┐
24:  x = ─────── + ───── + ─ + î │ ───────── - ──────── │
           4         4       2        └    4           4      ┘


         1/3       2/3                   1/6         1/6
        5         5        1        ┌  675       16875    ┐
25:  x = ─────── + ───── + ─ + î │ ─────── - ───────── │
           4         4       2        └   4           4      ┘
```

Therefore, the real value from expression #23, $x = 5^{1/3}/2 - 5^{2/3}/2 + 1/2$, is an inflection point.

The analysis of $h(x)$ proceeds in a similar manner. Author

```
DIF(x lnx+sinx,x)
```

and Simplify to obtain

```
         d
26:  ── (x LN (x) + SIN (x))
        dx

27:  LN (x) + COS (x) + 1
```

Now, we should be able to get the critical points by executing soLve. However, Derive cannot solve this transcendental expression exactly, so we will help by changing the mode with Options Precision Approximate. In this mode, the soLve command performs the bisection algorithm. Before we execute the command, we need to know a reasonable interval to search. Let's graph the function to obtain that interval. Simply execute `Plot Overlay Plot` to obtain the following plot:

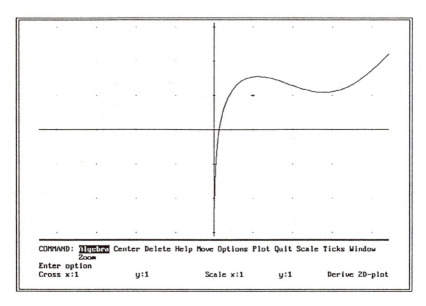

```
COMMAND: Algebra Center Delete Help Move Options Plot Quit Scale Ticks Window
         Zoom
Enter option
Cross x:1              y:1            Scale x:1      y:1       Derive 2D-plot
```

The function crosses the x-axis between 0 and 1. Now return to the `Algebra` window and reissue `soLve`. This time the bounds for the bisection algorithm are requested. Use the value of 0 for the lower bound and 1 for the upper bound to obtain

```
27:  LN (x) + COS (x) + 1

28:  x = 0.136601
```

The second derivative for this function is found by highlighting the first derivative and executing `Calculus Differentiate`. Then select `Simplify` to obtain

```
      d
29:   ──  (LN (x) + COS (x) + 1)
      dx


      1
30:   ──  - SIN (x)
      x
```

Use the **Manage Substitute** command to evaluate the second derivative at the critical value. The positive result determines that $x = 0.136601$ is the value where the function attains a local minimum.

```
          1
31:   ──────────  - SIN (0.136601)
       0.136601

32:   7.18441
```

Example 2.4: Integration

The union of the mathematician with the poet, fervor with measure,
passion with correctness, this surely is the ideal.

—William James [1879]

Subject: Evaluation of Definite and Indefinite Integrals

References: Sections 1.7 and 1.11

Problem: Compute the following integrals:

$$\int x^2 e^x \, dx,$$

$$\int_0^2 \ln(x^2 + 1) \, dx,$$

$$\int_0^{2\pi} \cos(x^2) \, dx.$$

Approximate the integral (area) in the interval $0 < x < 1$ of the function represented by the following table of data points:

x	y
0.0	3
0.2	4
0.4	5
0.5	2
0.6	2.1
0.9	3
1.0	2.9

Solution: Integrals are set up with the `Calculus Integrate` menu command or with the in-line `INT` command. Using the first method for the first integral, Author `x^2  Alt-e^x`. Then select `Calculus Integrate`. Answer the menu queries as follows: variable of integration: x; lower limit: (*leave blank*); upper limit: (*leave blank*).

The blank responses for the limits indicate this is indefinite integration or antidifferentiation. Execute `Simplify`. The set-up integral and the result of the integration are as follows:

$$2: \quad \int x^2\, \hat{e}^x\, dx$$

$$3: \quad \hat{e}^x\, (x^2 - 2x + 2)$$

Notice that the answer does not contain an arbitrary constant of integration. Derive does not automatically add in the constant. Let's check ourselves and Derive by taking the derivative of this expression to see if it is equivalent to the integrand. Just select `Calculus Differentiate` and respond with differentiation with respect to x and an order of 1. `Simplify` to obtain the result we desired.

$$2: \quad \int x^2\, \hat{e}^x\, dx$$

$$3: \quad \hat{e}^x\, (x^2 - 2x + 2)$$

$$4: \quad \frac{d}{dx}\, \hat{e}^x\, (x^2 - 2x + 2)$$

$$5: \quad x^2\, \hat{e}^x$$

We will evaluate the second integral

$$\int_0^2 \ln(x^2 + 1)\, dx$$

with the second method for integration discussed above by using the in-line `INT` command. Author

`INT(ln(x^2+1),x,0,2)` .

This is a definite integral, so the limits of integration are the third and fourth arguments. If this had been an indefinite integral, these parameters would have been omitted. Simplify to obtain the following result:

$$
6: \quad \int_{0}^{2} LN\ (x^{2} + 1)\ dx
$$

$$
7: \quad -2\ ATAN\ \left[\frac{1}{2}\right] + 2\ LN\ (5) + \pi - 4
$$

This is the exact solution found by symbolic integration. To obtain an approximation to these real numbers, select the approX menu command, which produces the following decimal approximation:

$$
7: \quad -2\ ATAN\ \left[\frac{1}{2}\right] + 2\ LN\ (5) + \pi - 4
$$

$$
8: \quad 1.43317
$$

Of course, the third integral

$$
\int_{0}^{2\pi} \cos(x^2)\, dx
$$

can be set up using either method. The in-line set-up command is *Authored* by

INT(cos(x^2),x,0,2pi) .

This time the Simplify command doesn't help. Derive just returns the original set-up for the integral.

$$9: \quad \int_{0}^{2 \pi} \cos(x^2) \, dx$$

$$10: \quad \int_{0}^{2 \pi} \cos(x^2) \, dx$$

The reason for this is that Derive cannot integrate this integrand exactly. However, Derive can still help. Execute approX to obtain an adaptive quadrature approximation to the definite integral. The approximate value is as shown in the following display:

$$10: \quad \int_{0}^{2 \pi} \cos(x^2) \, dx$$

$$11: \quad 0.704683$$

The accuracy goal in the form of significant digits for this numerical approximation is controlled by the Options Precision menu command. Execute the command and enter the number of digits of accuracy desired. Thereafter, Derive tries to achieve that level of accuracy by approximating the error in the numerical technique. If Derive is doubtful that it obtained the requested accuracy, it gives a warning of "Dubious Accuracy."

In order to approximate integrals of functions represented by discrete data using the trapezoidal rule, the INT_DATA command from the utility file NUMERIC is used. To load this file, execute Transfer Load Utility for NUMERIC. To use the command, the data must be in a matrix with 2 columns.

For our data file

x	y
0.0	3
0.2	4
0.4	5
0.5	2
0.6	2.1
0.9	3
1.0	2.9

execute Declare Matrix with 7 rows and 2 columns. Type in the data values for the appropriate entries. The result is

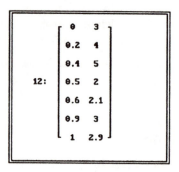

The Derive statement #12 can then be used as the argument of the INT_DATA command. To do this, Author INT_DATA #12 . Then execute the

`Simplify` command to obtain the matrix.

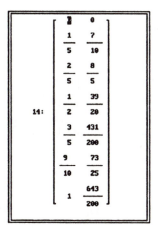

$$14: \begin{bmatrix} 3 & 0 \\ \dfrac{1}{5} & \dfrac{7}{10} \\ \dfrac{2}{5} & \dfrac{8}{5} \\ \dfrac{1}{2} & \dfrac{39}{20} \\ \dfrac{3}{5} & \dfrac{431}{200} \\ \dfrac{9}{10} & \dfrac{73}{25} \\ 1 & \dfrac{643}{200} \end{bmatrix}$$

The matrix shows the approximate antiderivative function for the x points in the data set. For the integral over the entire interval $0 < x < 1$, the last result is used. Therefore, the approximation to the integral or area for our problem is $643/200$.

Exercise 2.5: Area Between Graphs of Functions

I have made such wonderful discoveries that I am myself lost in astonishment.

—Janos Bolyai [1823]

Subject: Finding the Area Between Graphs of Two Functions

Purpose: To use integration techniques to find the area between the graphs of two functions.

References: Sections 1.7, 1.10, 1.12, and 2.4

Given: The two polynomial functions are defined by

$$f(x) = -4x^2 + x + 3,$$
$$g(x) = x^4 - 3.$$

Exercises:

1. Plot the two functions and find their intersection points.

2. Find the area of the region between the graphs of two functions interior to the two intersection points.

3. What is the centroid of this region?

4. Find the area of the region formed by the graph of $f(x)$ above the x-axis.

5. Find the area of the region formed by the graph of $g(x)$ below the x-axis.

6. What are the maximum vertical and horizontal distances of the region described in #2?

Example 2.6: Series

The eternal silence of these infinite places terrifies me.

—Blaise Pascal [1670]

Subject: Summations and Infinite Series

Reference: Section 1.7

Problem: Evaluate the following summations and series:

$$S_1 = \sum_{k=1}^{10} \frac{1}{k},$$

$$S_2 = \sum_{k=1}^{100} \frac{1}{k},$$

$$S_3(n) = \sum_{k=1}^{n} (2k - 1),$$

$$S_4 = \sum_{k=0}^{\infty} 4(0.8)^k,$$

$$S_5 = \sum_{k=1}^{\infty} \frac{1}{k(k+1)},$$

$$S_6 = \sum_{k=1}^{\infty} \frac{2}{10^k}.$$

Solution: There are two ways in Derive to set up summations. The first uses the `Calculus Sum` menu command and answers to the queries of the submenu. The second uses the `SUM` in-line function with the appropriate arguments.

Using the first method for S_1, `Author` the expression $\boxed{1/k}$. Then, select `Calculus Sum` and enter k for the summation variable, 1 for the lower limit, and 10 for the upper limit. `Simplify` to obtain

$$1: \quad \frac{1}{k}$$

$$2: \quad \sum_{k=1}^{10} \frac{1}{k}$$

$$3: \quad \frac{7381}{2520}$$

The Derive in-line command to set up the summation for S_2 is

$$\boxed{\texttt{SUM(1/k,k,1,100)}} .$$

Author this command and approX this time to see a 6-digit decimal approximation. The results are

$$4: \quad \sum_{k=1}^{100} \frac{1}{k}$$

$$5: \quad \frac{14466636279520351160221518043104131447711}{27888150091884990865813523574124921422272}$$

$$6: \quad 5.18737$$

For $S_3(n)$, Author

$$\boxed{\texttt{SUM(2k-1,k,1,n)}} .$$

Simplify to get the functional expression

$$7: \quad \sum_{k=1}^{n} (2k - 1)$$

$$8: \quad n^2$$

There is no problem in setting up an infinite series. Simply use `inf`, Derive's keystrokes for ∞, for the upper limit in either the in-line or menu command. Therefore, the in-line command for S_4 is

$$\boxed{\text{SUM(4(0.8)\textasciicircum k,k,0,inf)}}\ .$$

`Author` this expression and this time execute `approX` to obtain

```
         ∞      k
9:       Σ   4 0.8
        k=0

10:   20
```

Set up the summation for S_5 using one of the two methods and `Simplify` to obtain

```
         ∞        1
11:      Σ     ─────────
        k=1   k (k + 1)

12:   1
```

The command and its result upon issuing `approX` for S_6 are as follows:

```
              ∞      2
              Σ    ─────
13:     k=1      k
                   10

14:    0.222222
```

Exercise 2.7: Bouncing Ball

I always think it better, whenever possible, not to begin at the beginning, as it is always the most difficult part.

—Schiller [1796]

Subject: How Far Does a Bouncing Ball Travel?

Purpose: To use summations and properties of series to determine the distance travelled by a bouncing ball.

References: Sections 1.7, 1.10, and 2.6

Given: You drop a ball from h meters above a hard, flat surface. Each time the ball bounces, it rebounds a factor of r times its previous height.

Exercises:

1. If h is 10 and after the first bounce the ball rebounds to 3.7 meters, find r and the height of the rebound after 10 bounces.

2. How far did the ball travel in the first 10 bounces (including the rebound following the 10th bounce)?

3. Graph the rebound height for these first 10 bounces.

4. With r and h from #1, find the total distance the ball will travel until it stops. Does this make sense?

5. If h is doubled to 20, how does the total distance change?

6. If r is halved (from #1), how does the total distance change?

7. If $h = 10$, what factor r produces an equal travel distance for the first 2 bounces and all the remaining bounces thereafter?

8. Ask yourself other questions related to this problem that you would like to investigate. Determine if Derive is a proper tool to use in this investigation by trying to answer your questions.

3

First-Order Differential Equations

There are things which seem incredible to most men who have not studied mathematics.

—Archimedes

Example problems involving first-order differential equations have been worked out to show the power of Derive as a problem-solving tool in this subject. By reading these problems and working along with Derive, you should get a better feel for the subject of differential equations. Some of these examples also show the limitations of Derive, and help to show when to use Derive and the computer and when not to. Some examples involve models of applications, while others are posed in a purely mathematical context. The problems are similar to those typically found in undergraduate applied differential equations textbooks.

The exercises are to be done by you. These problems direct you to learn new things about first-order differential equations and Derive. Some of the problems, solution techniques, and Derive commands are similar to those in the example problems and sometimes refer to those examples as a source of help. Some of the exercises in this chapter lead you to explore the mathematics and software and to discover new results as you would in a laboratory course.

Example 3.1: Separable Differential Equation

There is nothing so captivating as new knowledge.

—Peter Latham [1870]

Subject: Solving the Logistics Equation, a Separable Differential Equation, Using Utility File ODE1

References: Sections 1.10 and 1.13

Problem: Find the solution to the modified logistics equation

$$\frac{dP}{dt} = P(a - bP)(1 - cP^{-1})$$

with a, b, $c > 0$ and $P(t) = 1000$ at $t = 0$.

Solution: This is a separable differential equation; therefore, the function `Separable` in file ODE1 can be used to solve the equation. The utility file ODE1 is loaded using the `Transfer Load Utility` command and giving $\boxed{\text{ODE1}}$ for the file name.

If the differential equation is in the form $y' = p(x)q(y)$ with $y(x_0) = y_0$, then the `SEPARABLE` function can be used to solve the equation. The function is entered as

$$\boxed{\texttt{SEPARABLE(p(x),q(y),x,y,x0,y0)}} \ .$$

Before this command is entered for our example, the given conditions on the parameters a, b, and c must be established. This is done with the `Declare Variable` command, making a, b, and c all `Positive`. This has to be done one variable at a time. The keystrokes **d v** result in the following menu display:

```
DECLARE VARIABLE name: _

Enter name or type "default"
User                          Free:100%              Derive Algebra
```

Next, we need to declare a as positive. The sequence of screen displays showing the Derive inputs needed to declare the variable a positive are as follows:

```
DECLARE VARIABLE name: a

Enter name or type "default"
User                              Free:100%              Derive Algebra
```

```
DECLARE VARIABLE: [Domain] Value

Enter option
User                              Free:100%              Derive Algebra
```

```
DECLARE VARIABLE DOMAIN: [Positive] Nonnegative Real Complex Interval

Select domain of a
User                              Free:100%              Derive Algebra
```

For this differential equation, the independent variable is t (instead of x) and the dependent variable is P (instead of y). Placing this equation in proper form produces the functions $p(t) = 1$ and $q(P) = P(a - bP)(1 - cP^{-1})$. Now, to solve this equation, execute the Author command and enter

$$\boxed{\text{SEPARABLE}(1,P(a-bP)(1-c/P),t,P,0,1000)} \; .$$

The command line and working area show this as

```
AUTHOR expression: separable(1,p(a-bp)(1-c/p),t,p,0,1000)_

Enter expression
User                              Free:98%               Derive Algebra
```

$$27: \quad \text{SEPARABLE} \left[1, \; p \; (a - b \; p) \left[1 - \frac{c}{p} \right], \; t, \; p, \; 0, \; 1000 \right]$$

The next step is to Simplify this expression to obtain the implicit solution. The resulting Derive statement is as shown.

$$28: \quad \frac{\text{LN } (1000\ b - a)}{a - b\,c} + \frac{\text{LN } (b\ p - a)}{b\,c - a} + \frac{\text{LN } (1000 - c)}{b\,c - a} + \frac{\text{LN } (p - c)}{a - b\,c} = t$$

In order to obtain an explicit form for $P(t)$, ask Derive to soLve for P. The result is

$$29: \quad p = \frac{a\,e^{a\,t}\,(c - 1000) + c\,e^{b\,c\,t}\,(1000\ b - a)}{b\,e^{a\,t}\,(c - 1000) + e^{b\,c\,t}\,(1000\ b - a)}$$

To check the initial condition, highlight the expression and execute Manage Substitute. Enter the value 0 for the variable t. Just press **Enter** when queried for replacement of the other unknowns. The result is

$$30: \quad p = \frac{a\,e^{a\,0}\,(c - 1000) + c\,e^{b\,c\,0}\,(1000\ b - a)}{b\,e^{a\,0}\,(c - 1000) + e^{b\,c\,0}\,(1000\ b - a)}$$

Now, Simplify this expression to achieve the given initial value of 1000. To check the solution, Author and enter the operator shown in the following input line:

```
AUTHOR expression: dif(p,t)-p(a-bp)(1-c/p)_

Enter expression
Simp(35)                    D:ODE1.MTH              Free:86%
```

Then, Manage Substitute the expression for the solution P (for this example, this is the right-hand side of expression #29) into the operator using the highlight and F3 key (be sure to delete the p in the input line before pressing F3). Simplify to get the following screen, which verifies the result:

$$38: \quad \frac{d}{dt} \; \frac{a\,e^{a\,t}\,(c-1000) - c\,e^{b\,c\,t}\,(a-1000\,b)}{b\,e^{a\,t}\,(c-1000) + e^{b\,c\,t}\,(1000\,b - a)} \; - \; \frac{a\,e^{a\,t}\,(c-1000) - c\,e^{b}}{b\,e^{a\,t}\,(c-1000) + e^{b\,c}}$$

$$39: \quad 0$$

Let's use some specific values for a, b, and c so we can plot one of the solution curves. We choose $a = 1$, $b = 0.5$, and $c = 0.5$. Manage Substitute these values into the solution (expression #29) and Simplify to get

$$40: \quad p = \frac{1\,e^{1\,t}\,(0.5 - 1000) - 0.5\,e^{0.5\,0.5\,t}\,(1 - 1000\,0.5)}{0.5\,e^{1\,t}\,(0.5 - 1000) + e^{0.5\,0.5\,t}\,(1000\,0.5 - 1)}$$

$$41: \quad p = \frac{2\,(1999\,e^{3\,t\,/\,4} - 499)}{1999\,e^{3\,t\,/\,4} - 1996}$$

Plot Plot this solution using the default scale of Derive to show the solution graphically with these parameter values we have determined. The result is as follows:

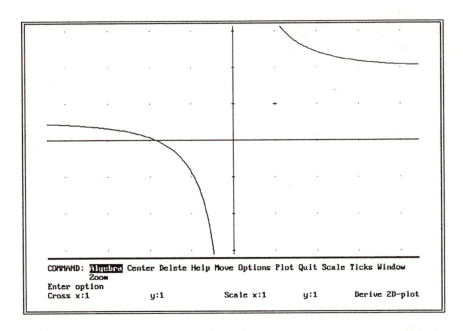

COMMAND: **Algebra** Center Delete Help Move Options Plot Quit Scale Ticks Window
         Zoom
Enter option
Cross x:1   y:1   Scale x:1   y:1   Derive 2D-plot

This plot does not show much of the overall behavior of the solution. This is because the default plotting window shows approximately $-4.5 < x < 4.5$ and $-3 < y < 3$. Let's change the window size by executing Scale and setting x-scale $= 1$ and y-scale $= 100$. The new plot is automatically redrawn to the new scale as follows:

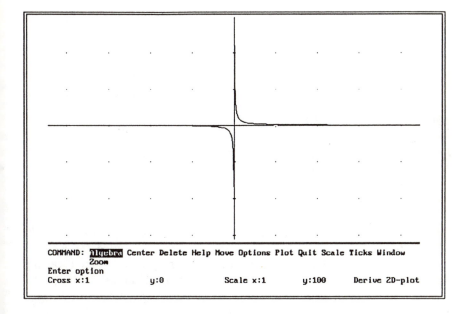

Exercise 3.2: Population Growth

> *Population, when unchecked, increases in a geometrical ratio. Subsistence increases only in an arithmetical ratio. A slight acquaintance with numbers will show the immensity of the first power in comparison of the second.*

> —Thomas Malthus [1798]

Subject: Analyzing and Determining Population Growth

Purpose: To analyze a model for population growth of a species in a limited-resource environment.

References: Sections 1.13 and 3.1

Given: There are several differential equations used for population growth. We will use a model for a limited-resource environment called the logistics equation. In this model, P is the population size and t is time from a starting date measured in years. The differential equation is written as

$$\frac{dP}{dt} = r(M - P)P$$

with $r, M, P \geq 0$.

Exercises:

1. Plot the phase plane curves (dP/dt versus P) for the model with parameter values $r = 0.5$ and $M = 500$. (**Hint:** For this plot, use y for dP/dt and x for P; use x-scale ≈ 200 and y-scale $\approx 100{,}000$.)

2. Use the cross on the plot screen to approximate the equilibrium points for the model, and then classify these points as stable or unstable.

3. If $P(0) = 600$ and using the parameters given in #1, solve the differential equation using the operation SEPARABLE in the utility file ODE1.MTH.

4. Begin a new plot screen either by deleting the stability curves or by overlaying a new screen, and plot the solution obtained in #3.

5. Ask yourself other questions related to this problem that you would like to investigate. Determine if Derive is a proper tool to use in this investigation by trying to answer your questions.

Example 3.3: Exact Differential Equation

A collection of facts is no more a science than a heap of stones is a house.

—Jules Henri Poincaré [1903]

Subject: Solving an Exact Differential Equation Using Utility File ODE1

References: Sections 1.10 and 1.13

Problem: Find the particular solution of

$$(y^2 \cos x - 3x^2 y - e^x - 1) + (2y \sin x - x^3 + \ln y + 2)y' = 0,$$

with $y = 1$ at $x = 0$.

Solution: Exact differential equations are often easily solved by integration. In Derive, the command **EXACT** in file ODE1 can be used to check for exactness and to solve the equation. The utility file ODE1 is loaded using the **Transfer Load Utility** command and giving $\boxed{\text{ODE1}}$ for the file name.

This differential equation is already in the proper form for the **EXACT** command of $p(x,y) + q(x,y)y' = 0$ with initial condition $y(x_0) = y_0$. Therefore, the form of the solution command is

$$\boxed{\texttt{EXACT(p(x,y),q(x,y),x,y,x0,y0)}} \; .$$

The equation is exact if a solution is provided. If the equation is not exact, "`inapplicable`" is returned. The input command to solve this problem is

$$\boxed{\texttt{EXACT(y\textasciicircum 2cosx-3x\textasciicircum 2y-Alt-e\textasciicircum x-1,2y sinx-x\textasciicircum 3+lny+2,x,y,0,1)}} \; .$$

Simplify this expression to obtain the solution. The command and its result are shown below.

```
           2            2     x               3
29:  EXACT (y  COS (x) - 3 x  y - ê  - 1, 2 y SIN (x) - x  + LN (y) + 2, x, y, 0

        x            2        3
30:  - ê  + y LN (y) + y  SIN (x) - x  y - x + y = 0
```

We now try to solve explicitly for y using the **soLve** command. However, the result shows that there is a problem. The unchanged expression implies that Derive cannot help. This implicit equation is transcendental in y, so it is impossible to determine an explicit solution for y in terms of Derive's built-in functions. This equation is also transcendental in x, so Derive can't solve for x in terms of y either. This is unfortunate since Derive Version 2 does not plot implicit functions. However, don't despair; we do have the implicit solution to the differential equation because of Derive's help.

In fact, we may be able to plot a few explicit solution points. First, set the Options Precision command to Approximate. Then use the Manage Substitute command to replace x with specific values. Then execute the soLve command to find y via the bisection method. The search bounds for the method may have to be adjusted for different values of x. The results of these steps for $x = 1$, 2, 2.5, and 3 are as follows:

$$32: \quad -8^1 + y \, LN\,(y) + y^2 \, SIN\,(1) - 1^3 \, y - 1 + y = 0$$

$$33: \quad y = 1.78563$$

$$34: \quad -8^2 + y \, LN\,(y) + y^2 \, SIN\,(2) - 2^3 \, y - 2 + y = 0$$

$$35: \quad y = 7.02438$$

$$36: \quad -8^{2.5} + y \, LN\,(y) + y^2 \, SIN\,(2.5) - 2.5^3 \, y - 2.5 + y = 0$$

$$37: \quad y = 20.5764$$

$$38: \quad -8^3 + y \, LN\,(y) + y^2 \, SIN\,(3) - 3^3 \, y - 3 + y = 0$$

$$39: \quad y = 149.834$$

These points, along with the point of the initial condition, can be placed in an array and plotted. Author

$$[[0,1], \ [1,1.785], \ [2,7.024], \ [2.5,20.576], \ [3,149.834]] \ .$$

Then select the Plot menu and Overlay location. Change the Scale to $x = 1$ and $y = 50$, and set Options State Mode to Connected. Then execute Plot. The resulting plot is

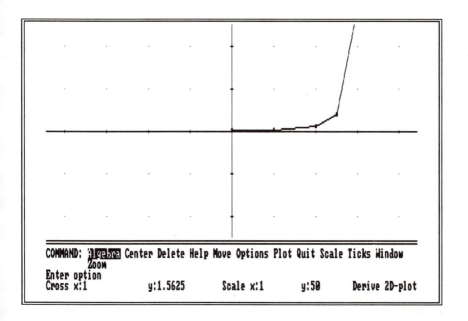

```
COMMAND: ]Algebra Center Delete Help Move Options Plot Quit Scale Ticks Window
          Zoom
Enter option
Cross x:1              y:1.5625        Scale x:1        y:50        Derive 2D-plot
```

There are other ways to produce solution points for this differential equation. One possibility would be to use a numerical procedure to approximate the solution. See Chapter 4 for the use of numerical procedures.

Another more general and powerful command is available in utility file ODE1 to solve first-order equations. The command DSOLVE1 can solve equations that are exact, linear, separable, or homogeneous. This command assumes the equations are in the same form as EXACT, $p(x,y) + q(x,y)y' = 0$ with condition $y(x_0) = y_0$. The arguments for DSOLVE1 are

$$\boxed{\texttt{DSOLVE1(p(x,y),q(x,y),x,y,x0,y0)}}\;.$$

If the equation cannot be solved with this method, "inapplicable" is returned. The expression to **Author** to solve this problem using DSOLVE1 is

$$\boxed{\texttt{DSOLVE1(y\^{}2cosx-3x\^{}2y-Alt-e\^{}x-1,2y sinx-x\^{}3+lny+2,x,y,0,1)}}\;.$$

This expression can be entered without retyping the entire expression. First, select **Author**. Then highlight the entire expression previously entered that used the EXACT command and press the **F3** key to bring that expression into the author line. Change EXACT to DSOLVE1 by moving the cursor using the **Ctrl-S** and **Ctrl-D** keys and replacing and inserting the needed letters (see Section 1.3 for more information). Press **Enter** and then **Simplify** this expression to obtain the same implicit solution given by the EXACT command.

The command and its result are shown below.

41: DSOLVE1 (y^2 COS (x) - 3 x^2 y - θ^x - 1, 2 y SIN (x) - x^3 + LN (y) + 2, x, y,

42: - θ^x + y LN (y) + y^2 SIN (x) - x^3 y - x + y = 0

Another tool available in Derive to analyze solutions is a direction field. Its function definition is in utility file ODE_APPR. To obtain a plot of the direction field for this equation in the region $0 \leq x \leq 4$ and $0 \leq y \leq 50$, Transfer Load Utility the file ODE_APPR, then Author and approX the combined expression given in the following two boxes:

DIRECTION_FIELD(-(y^2cosx-3x^2y-**Alt**-e^x-1)/(2y sinx-x^3+lny+2) ,

x,0,4,8,y,0,50,8) .

Then plot the resulting vector with the x-scale set at 0.5, the y-scale set at 15, the plot's center at $(2,25)$, and the Options State set to Connected Mode and Small Size. The resulting plot is shown.

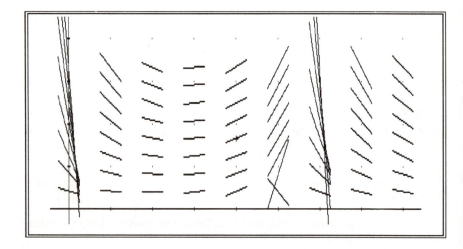

Example 3.4: Radioactive Decay

These methods of learning about nature are increasingly more important in more and more fields. They also underlie the process by which engineers create the technologies that exercise vast influence over all our lives.

—Samuel Goldberg, *UME Trends* [1990]

Subject: Using Radioactive Decay of Carbon-14 to Determine the Age of Paintings

References: Sections 1.11 and 1.13

Problem: Carbon-14 has a half-life of about 5700 years; i.e., in this time 50% of carbon-14 decays into nitrogen. The rate of decay of a radioactive element like carbon-14 is proportional to the amount of the element present. If $a(t)$ is the amount of carbon-14 at t years, the model that describes its decay is

$$\frac{da}{dt} = -ka,$$

with $a(0) = c$.

If a painting has between 90% and 91% of its original carbon-14, how many years ago was the painting created?

Solution: The proportionality constant k is not known. However, from the half-life information, we know $a(5700) = 0.5a(0) = 0.5c$. By rewriting the equation in the form

$$\frac{da}{dt} + ka = 0,$$

the equation is linear because it fits the general form $y' + p(x)y = q(x)$ with the independent variable t (instead of x) and dependent variable a (instead of y). Initially, k and c are unknown constants. The appropriate arguments for the LINEAR1 command can be determined from the form LINEAR1(p,q,x,y,x0,y0). For more information on this command, see Section 1.13.

To use this command for this problem, **Transfer Load Utility** the utility file ODE1. Then **Author** the expression

LINEAR1(k,0,t,a,5700,0.5c) .

Execute Simplify to obtain

```
27:   LINEAR1 (k, 0, t, a, 5700, 0.5 c)

                  k (5700 - t)
              c ê
28:   a = ────────────────────────
                      2
```

Now substitute the initial condition $a(0) = c$ into the solution. Use the Manage Substitute command to do this. We want to substitute 0 for t and c for a and leave the unknowns c and k as they are. Just press **Enter** when Derive asks what to substitute for c and k. The result is

```
                         k (5700 - t)
                     c ê
28:   a = ────────────────────────
                      2

                         k (5700 - 0)
                     c ê
29:   c = ────────────────────────
                      2
```

We want to find the expression for k in terms of c, so we execute the soLve command for the solve variable k. The resulting output expression is what we wanted.

$$29: \quad c = \frac{c\, \hat{e}^{k\,(5700 - 0)}}{2}$$

$$30: \quad k = \frac{LN\,(2)}{5700}$$

Now that we know the value of k for carbon-14, we can find the values of t where $a = 0.90c$ and $a = 0.91c$. Highlight the solution function (expression #28) and use the Manage Substitute command to replace k with its value $\ln(2)/5700$. Then Simplify to obtain the equation

$$34: \quad a = \frac{c\, \hat{e}^{LN\,(2)\,/\,5700\,(5700 - t)}}{2}$$

$$35: \quad a = 2^{-t/5700}\, c$$

Select Manage Substitute for this expression to replace a with 0.9 and c with 1. Try executing soLve with t being the solve variable. Derive does not seem to be able to solve the equation. The system keeps working, but no progress is being made. In this case, abort the processing by pressing **Esc**. This time, let's approximate a solution using the bisection method. To start this method, we need to know the end points of an interval that includes the solution. In this case, -1.21605×10^{-4} is small, so t must be quite large (near 10^4) to satisfy the equation. Let's try the interval $500 < t < 2000$. If our interval is not appropriate, Derive will indicate this by printing No solution found. Another method of estimating the solution would be to graph the function. In this case, we won't produce the graph.

To implement the bisection algorithm, we issue Options Precision Approximate. The required digits of accuracy is set to 6. Execute soLve with the parameters Lower: 500 and Upper: 2000. The **Tab** key moves the cursor between these two entry fields. The result of the approximation is

```
                    - t/5700
    36:   0.9 = 2              1

    37:   t = 866.416
```

The same method is used for $a = 0.91c$. The resulting display is as follows:

```
                    - t/5700
    38:   0.91 = 2             1

    39:   t = 775.548
```

We have determined the painting is between 775 and 866 years old.

Once again, the more general DSOLVE1 command could be used to solve this equation. The command is slightly different from that of the LINEAR1 command since the equation must be placed in the form $p(x, y) + q(x, y)y' = 0$ to identify the command's arguments. For this problem, Author

```
DSOLVE1(ka,1,t,a,5700,0.5c)
```
.

Execute Simplify to obtain the same solution as that produced using the LINEAR1 command.

```
    40:  DSOLVE1 (k a, 1, t, a, 5700, 0.5 c)

    41:  LN (2 a) - LN (c) + k (t - 5700) = 0

                    k (5700 - t)
               c e
    42:  a = ─────────────────────
                     2
```

Exercise 3.5: Linear Equations

> ... *the different branches of Arithmetic—Ambition, Distraction, Uglification, and Derision* ...

> —Lewis Carroll, *Alice in Wonderland* [1865]

Subject: Solving Linear and Bernoulli Equations

Purpose: To solve and analyze linear and Bernoulli equations.

References: Sections 1.6, 1.7, and 1.13

Given: The commands LINEAR1 and BERNOULLI are provided in the utility file ODE1.MTH to solve initial-value problems of special forms. There are descriptions of these commands on-line in the Help menu for file ODE1.MTH.

Exercises:

1. Solve and plot the solution of

$$xy' + (1 + x)y = e^{-x}\sin(2x), \quad y(1) = \sqrt{2}.$$

What is the limiting value of the solution as $x \to \infty$?

2. Solve and plot the solution of

$$y' + (\tan x)y = \cos^2 x, \quad y(\pi) = 2.$$

3. Solve and plot on the same axes the solutions of

$$y' + y = xy^n, \quad y(0) = 1,$$

for $n = 2$ and 3.

4. Solve and plot on the same axes the solutions of

$$xy' + y = 2x, \quad y(a) = b,$$

for the following four pairs of values for a and b: $(a, b) = (-2, 0)$, $(-1, -2)$, $(1, 2)$, $(2, 0)$.

Exercise 3.6: Waste Disposal

Science can tell us how to do many things, but it cannot tell us what ought to be done.

—Anonymous

Subject: A Model for Waste Disposal in the Ocean

Purpose: To analyze a plan for dumping toxic waste in the ocean using a first-order, linear differential equation.

References: Sections 1.7 and 1.13

Given: One method proposed for disposing of toxic waste is to place it in sealed drums and dump the drums in a deep part of the ocean. Tests showed that drums could be made that would never leak from corrosion, but there still was concern over their breaking open from impact with the ocean floor. Through the summation of the forces of gravity, buoyancy, and drag, a model for the velocity $(v(t))$ of the drum's descent in time through the water is developed as

$$\frac{dv}{dt} + \frac{ag}{wz}v = \frac{g}{w}(w - b),$$

where a is the coefficient of drag for the drums, g is the acceleration of gravity (32 ft/sec^2), w is the density of the toxic waste, b is the density of the water (produces buoyancy), and z is the volume of the drum.

Exercises:

1. Solve the differential-equation model using two different commands in file ODE1.MTH. What is the maximum velocity of the drums?

2. If 55-gallon drums (7.4 ft^3) are dumped from rest by rolling them off the side of a barge, the waste has density 80 lbs/ft^3, water has density 62.5 lbs/ft^3, and the drag coefficient for this style drum is 0.10, plot the velocity for $0 < t < 30$. What is the maximum velocity of the drums for these values of the parameters?

3. Integrate the function in #1 for velocity to obtain an expression for the distance traveled (y) by the drums in terms of time t.

4. If the parameter values are as given in #2 and if the ocean depth is 400 feet, how fast are the drums traveling at impact?

5. Does it reduce the speed at impact if smaller drums are used? (Assume the drag coefficient is unchanged.) Explain.

6. If the maximum safe impact velocity is 75 ft/sec and the other conditions are the same as in #2, what is the maximum ocean depth for safe dumping?

7. If the density of the waste is increased to 90 lbs/ft^3 and the other conditions are the same as in #6, what is the maximum ocean depth for safe dumping?

8. Ask yourself other questions related to this problem that you would like to investigate. Determine if Derive is a proper tool to use in this investigation by trying to answer your questions.

Exercise 3.7: Newton's Law of Cooling

I do not love . . . to be dunned and teezed by forreigners about mathematical things. . . .

—Isaac Newton [1699]

Subject: Modeling and Solving Newton's Law of Cooling

Purpose: To solve and analyze models for cooling of a space capsule after splashdown.

References: Sections 1.10 and 1.13

Given: The model for the cooling of the exterior of a space capsule is

$$\frac{dT}{dt} = -k(T - \beta), \quad t > 0, \quad T(0) = \alpha,$$

where T is the temperature (in degrees F) of the capsule, t is the time (in seconds) after splashdown, β is the temperature of the ocean, α is the initial temperature of the capsule, and k is a constant of proportionality. (**Hint:** In this case, both the independent and dependent variables are the same letter. If Derive is in the case-insensitive mode, it will treat T and t the same. Put Derive in the case-sensitive mode by executing the **Options Input** command. Press **Tab** and **S** to select the case-sensitive mode.)

Exercises:

1. Solve for the temperature T as a function of t and the parameters α, β, and k. (**Hint:** Consider using an operation from utility file ODE1.MTH, and be sure you have read the hint in the problem statement above.)

2. If the capsule temperature was 650° F at splashdown, the ocean is 40° F, and after 20 seconds the capsule temperature has dropped to 450° F, find the value of k.

3. Plot the solution (T versus t) in the above scenario. The domain of interest is $0 \leq t \leq 200$ seconds.

4. What changes are needed in the model if the capsule is lifted out of the water at $t = 100$ seconds and remains at an air temperature of 80° F?

5. Solve for T over the domain of interest $100 \leq t \leq 200$ in the above scenario (#4) with k as found in #2.

6. Plot this new solution (from #5) on the same axes as the previous solution plotted in #3. (**Hint:** It may be helpful to use the STEP function to plot the new function for $t > 100$.)

7. What is the temperature difference of the capsule at $t = 200$ seconds between the 2 different scenarios?

Example 3.8: Drug Doses

Eventually there will be, I hope, some people who will find it profitable to decipher this mess.

—Evariste Galois [1832]

Subject: Determining Drug Doses Using a Differential Equation

References: Sections 1.6, 1.10, and 1.13

Problem: The decrease in concentration of drugs in the bloodstream is proportional to the concentration of the drug. If the concentration in milligrams per milliliter of a drug at time t hours is represented by $D(t)$, then the differential-equation model is $D'(t) = -kD(t)$. For the specific drug being administered, we know that $k = 0.27$, the minimum effective concentration is 2.2 mg/ml, and the maximum safe concentration is 3.4 mg/ml. It is reasonable to assume that the blood absorbs the drug instantaneously.

Devise a schedule of safe and effective doses of this drug for a body of a small animal with 1000 ml of blood.

Solution: Let's start with the dose at $t = 0$. We want to administer the maximum safe dosage and let it decrease in concentration until it reaches the minimum safe concentration. Then we will administer enough of the drug to raise the concentration back up to the maximum.

Therefore, at $t = 0$ we will administer $(3.4)(1000) = 3400$ mg of the drug to obtain a concentration of $D(0) = 3.4$. Next we need to find the time t when $D(t)$ decreases to 2.2. We will use the solution operators for differential equations in utility file ODE1.MTH. Load the functions from this file into the work area by executing **Transfer Load Utility**. Type ODE1 for the file name.

By writing the equation and initial conditions as $D'(t) + 0.27D(t) = 0$ and $D(0) = 3.4$, we see the equation is in linear form. Linear equations of the form

$$y' + p(x)y = q(x),$$

with $y = y_0$ at $x = x_0$, are solved with the command

$$\boxed{\texttt{LINEAR1(p,q,x,y,x0,y0)}}\ .$$

To solve our differential equation, **Author**

$$\boxed{\texttt{LINEAR1(0.27,0,t,d,0,3.4)}}\ .$$

Simplify this expression to obtain the solution shown in the following display:

```
27:   LINEAR1 (0.27, 0, t, d, 0, 3.4)

                 - 27 t / 100
           17 ê
28:   d = ─────────────────
                  5
```

We need to find the time t when $D(t) = 3.4e^{-0.27t} = 2.2$. To do this, Author

$$3.4\ \text{Alt-e}^\wedge(-2.7t)=2.2$$

and execute soLve. Find the decimal solution by selecting approX.

```
                - 0.27 t
29:   3.4 ê              - 2.2

30:   t = 1.61228
```

By our stated plan, at $t = 1.6122$ hours, we administer 3400 mg − 2200 mg (or 1200 mg) of the drug to increase the drug to its maximum safe concentration. We now know our drug dose schedule—administer 1200 mg of the drug every 1.6 hours.

Derive has several ways to represent this schedule by a discontinuous function. We could probably use the STEP function or program the function using the IF command. Probably the easiest way to represent this function is through the use of the MOD function found in the MISC utility file. The MOD(a,b) command evaluates a modulo b, or the remainder of a/b. This is just what is needed since the drug schedule repeats itself every 1.6122 hours, so we only need to know how long since the last drug dose. In order to plot the cycles of this schedule, Author

$$3.4\ \text{Alt-e}^\wedge(-0.27\ \text{MOD}(t,1.6122))\ .$$

Then select the Plot menu and Overlay location. Move the cross to $x = 4$ and $y = 2$, and execute Center. Leave the x- and y-scales at 1. Execute Plot to obtain the following plot of the drug concentration (mg/ml) over time:

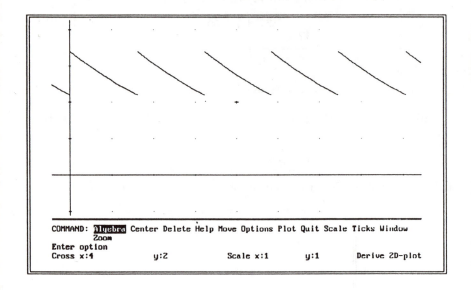

Notice how the concentration stays in the safe but effective region for all times.

Exercise 3.9: Oil Spill

The decision to seek the formulas that describe phenomena leads in turn to the question: What quantities should be related by formulas?

—Morris Kline [1985]

Subject: Modeling and Analyzing an Oil Spill

Purpose: To model oil flow with a first-order differential equation and to find the solution to the equation.

References: Sections 1.7 and 1.13

Given: The manufacturer of an oil-tanker ship is trying to reduce the damage caused by oil leakage in the event of a ruptured oil tank. The upper holding tank, which is above the water level in the company's newly designed ship, is assumed to be in the shape of a cylinder lying on its side. The shape and dimensions of the storage tank inside the ship are shown in the following sketch:

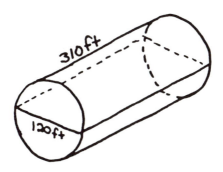

Exercises:

1. What is the capacity of this tank?

2. If an 8 square feet hole is opened in the bottom of the tank, determine the amount of time in seconds that it will take for the tank to empty. Assume that a form of Torricelli's Law applies so the velocity v of the oil through the hole is $v = 0.4\sqrt{2gh}$, where g is the acceleration due to gravity (32 ft/sec^2) and h is the height in feet of the oil above the hole.

3. What is the average rate of oil spillage?

4. How long does it take for half of the oil in the tank to spill? What is the average rate of spillage during this time?

5. How much oil has spilled after one hour (3600 seconds)?

6. Ask yourself other questions related to this problem that you would like to investigate. Determine if Derive is a proper tool to use in this investigation by trying to answer your questions.

Example 3.10: Solving First-Order Differential Equations

The study of Euler's works will remain the best school for the different fields of mathematics and nothing can replace it.

—Carl Friedrich Gauss

Subject: Solving a First-Order Homogeneous Differential Equation Using Utility File ODE1

References: Sections 1.10 and 1.13

Problem: Find the particular solution of the initial-value problem

$$y' = \frac{x^2 + y^2}{xy}$$

with $y(1) = -2$.

Solution: The most general solution command in utility file ODE1 is the DSOLVE1 command. In order to use this command, the equation must be in the general form of $p(x, y) + q(x, y)y' = 0$. Then the command is entered as

$$\boxed{\texttt{DSOLVE1(p,q,x,y,x0,y0)}}\ .$$

For this problem, the equation can be written as $-x^2 - y^2 + xyy' = 0$. Therefore, $p(x, y) = -x^2 - y^2$ and $q(x, y) = xy$. First, ODE1 is loaded using the **Transfer Load Utility** command and giving $\boxed{\texttt{ODE1}}$ for the file name. Then **Author** the expression

$$\boxed{\texttt{DSOLVE1(-x\textasciicircum2-y\textasciicircum2,xy,x,y,1,-2)}}\ .$$

```
            2   2
1:  DSOLVE1 (- x - y , x y, x, y, 1, -2)

                       2
                      y
2:  - LN (|x|) + ───────── - 2 = 0
                        2
                     2 x
```

This equation can be solved explicitly for y using the Derive soLve command and entering y as the solve variable. This procedure produces the following output:

3:   $y = - \sqrt{2} \times \sqrt{(\text{LN}\,(|x|) + 2)}$

4:   $y = \sqrt{2} \times \sqrt{(\text{LN}\,(|x|) + 2)}$

The negative expression in the solution is consistent with the initial condition. This equation is easily plotted using the Plot command with Overlay location. Once in the plot screen, use the Scale menu command to set the x-scale to 2 and the y-scale to 10. Remember from the table in Section 1.10 that this command sets the distance between each tick mark on each axis to the scale value entered. Then use the Move command to move the cross to $x = 3$ and $y = 0$. Execute the Center command to move the plot region so it is centered at the point (3,0). Ensure the proper function is highlighted back in the algebra window. Finally, issue the Plot command to plot the function. The resulting plot is shown.

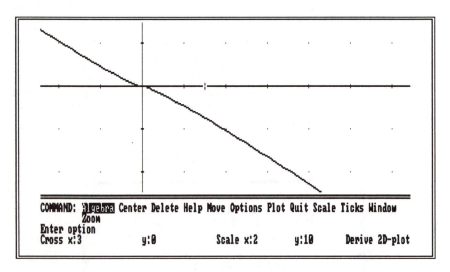

This equation may be solved using another method of Derive. Even though this equation is not exact or separable, it may be homogeneous. To check to see if the equation is homogeneous and to solve the equation using the homogeneous method, the function HOMOGENEOUS can be used. The differential equation needs to be in the proper form of $y' = r(x, y)$; therefore, Author the expression

$$\boxed{\text{HOMOGENEOUS}((x^2+y^2)/(xy),x,y,1,-2)}$$.

Simplify and soLve this expression to obtain the same solution that was produced using DSOLVE1.

If the equation had not been homogeneous, Derive would have responded to the command with "inapplicable."

4

Numerical Methods and Difference Equations

Yet today a tremendously powerful weapon of our own creation—namely mathematics—has given us knowledge and mastery of major areas of our physical world.

—Morris Kline [1985]

Example problems involving the numerical solution of first-order differential equations and difference equations have been worked out to show the power of Derive. By reading these problems and working along, you should get a better feel for these subjects. Some of the examples also show the limitations of Derive, and should help you to know when and when not to use Derive. Some of the examples involve applications, while others are posed in a mathematical context. The problems are similar to those found in undergraduate differential-equations textbooks.

The exercises are to be done by you. These problems give practice in problem solving. Some of the exercises in this chapter lead you to explore the mathematics and software and to discover new results on your own. As you solve the questions from these exercises, ask yourself additional questions and try to answer them. Good luck, and have fun learning and exploring.

Example 4.1: Picard Iteration

One of the charms of mathematics is that good mathematics never dies. It may fade from view, but it is not demolished by later discoveries.

—David Wells [1985]

Subject: Successive Iteration (Picard's Method)

References: Sections 1.10 and 1.13

Problem: Find an approximation to the solution of

$$y' = x^2 - 2y^2 - 1$$

with $y(0) = 0$.

Solution: This is a nonlinear differential equation that is not separable or exact and therefore does not lend itself easily to a simple closed-form solution. The approximation method of successive iteration can be used to approximate the solution for such problems. The method needs an initial approximation and the initial condition to get started. In this case, let's start with $y = x$ as the initial approximation to the solution. This approximation is the linear polynomial that satisfies the initial condition. Many other initial approximations are possible for this problem.

Derive's successive iteration operator is contained in the utility file ODE_APPR.MTH. Load this file into the work area first with the **Transfer Load Utility** command. The command PICARD(r(x,y),yprev,x,y,x0,y0) produces the next iterative approximation for the differential equation $y' = r(x,y)$ with approximation y_{prev} and initial condition $y(x_0) = y_0$. Therefore, for this problem the first iteration is performed by

```
PICARD(x^2-2y^2-1,x,x,y,0,0)
```

.

Simplify this operation to obtain the next approximation to the solution as shown.

```
            2     2
68:  PICARD (x  - 2 y  - 1, x, x, y, 0, 0)

          3
         x
69:  - ----- - x
          3
```

In entering the command for the next iteration, take advantage of the previous expressions by highlighting them and using the **F3** key to produce an input line and working display as shown.

```
AUTHOR expression: PICARD (x^2 - 2 y^2 - 1, - x^3 / 3 - x, x, y, 0, 0)

Enter expression
Simp(68)                D:2.17              Free:54%.           Derive Algebra
```

$$70: \quad PICARD \left[x^2 - 2 y^2 - 1, \ - \frac{x^3}{3} - x, \ x, \ y, \ 0, \ 0 \right]$$

Once again, the Simplify command produces the next successive approximation.

$$71: \quad - \frac{x (10 x^6 + 84 x^4 + 105 x^2 + 315)}{315}$$

Two more iterations produce the following display with some truncation:

$$72: \quad PICARD \left[x^2 - 2 y^2 - 1, \ - \frac{x (10 x^6 + 84 x^4 + 105 x^2 + 315)}{315}, \ x, \ y, \ 0, \ 0 \right]$$

$$73: \quad - \frac{x (5720 x^{14} + 110880 x^{12} + 714168 x^{10} + 2282280 x^8 + 7837830 x^6 + 113513}{42567525}$$

$$74: \quad PICARD \left[x^2 - 2 y^2 - 1, \ - \frac{x (5720 x^{14} + 110880 x^{12} + 714168 x^{10} + 2282280 x}{42} \right.$$

$$75: \quad - \frac{x (211466544432000 x^{30} + 8763001830704000 x^{28} + 151861523283500000 x^{26} +}{}$$

Plots of the second approximation and the last approximation to the solution are made using the Plot Overlay window. Scale the y-scale to 5 and execute Plot to obtain the following:

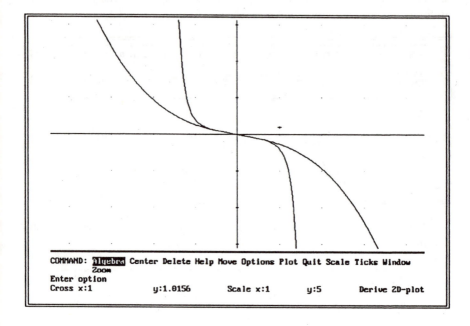

Example 4.2: Numerical Solution of Differential Equations

When you have eliminated the impossible, whatever remains, however improbable, must be the truth.

—Sir Arthur Conan Doyle, *The Sign of Four* [1890]

Subject: Numerical Solution of Differential Equations

References: Sections 1.10, 1.13, and 1.14

Problem: Given the following first-order, nonlinear differential equation

$$y' = \frac{y^2 \sin(2x)}{x^2},$$

with $y(1) = 2$, use the Euler and Runge-Kutta numerical methods to approximate solutions to the given equation with step-size $h = 0.15$ and number of steps $n = 50$.

Solution: The commands for these two numerical procedures are located in utility file ODE_APPR.MTH. Load this file through execution of **Transfer Load Utility** and by entering $\boxed{\texttt{ODE\_APPR}}$.

The command to perform the Euler method for a differential equation in the form $y' = r(x, y)$ is

$$\boxed{\texttt{EULER(r(x,y),x,y,x0,y0,h,n)}} \; .$$

Therefore, for this problem the Euler method is set up by the **Author** of the function

$$\boxed{\texttt{EULER(y\^{}2 sin(2x)/x\^{}2,x,y,1,2,0.15,50)}} \; .$$

In this command, an h value of 0.15 produces 50 solution values (since $n = 50$) for x values every 0.15 units apart, starting with the $x_0 = 1$. The iteration is performed by executing **approX**. The EULER function uses the Derive **ITERATE** command and, therefore, takes time and memory to execute. The result is a set of points (x, y), which approximate points on the solution curve. The command and the beginning of the set of points (truncated because of length) are as follows:

$$
15: \quad \text{EULER} \left[\frac{y^2 \, \text{SIN} \, (2 \, x)}{x^2}, \, x, \, y, \, 1, \, 2, \, 0.15, \, 50 \right]
$$

16:    [[1, 2], [1.15, 2.54557], [1.3, 3.09364], [1.45, 3.53154], [1.6, 3.74443],

Before we use the Plot menu to plot these points, we check the range
of the y values of the points. In order to see all 50 points, the **Ctrl-→** keys
move the visible area to the right. A quick look shows the range of y values is
$2 \le y \le 4$. Since we already know that $0 \le x \le 8.5$, we know the plot region
to produce using rescaling and centering. First we select Plot Overlay. Then
issue the Scale command and set the x-scale to 1.5 and the y-scale to 1. Move
the cross to $x = 4.5$ and $y = 2$, and execute Center. Then execute Plot to
get the following display:

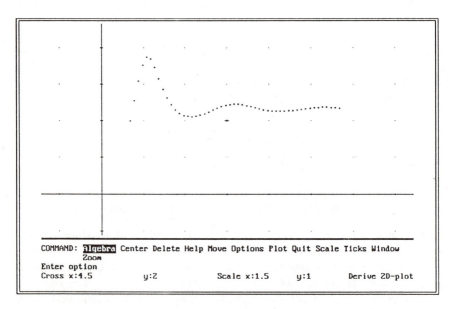

The Runge-Kutta method has a higher order of accuracy than the Euler
method. Of course, the cost of this accuracy is more computational work
per step. Let's see if this method improves the solution to this problem. The
command to perform the Runge-Kutta method is slightly different than the
command for the Euler method. This is because the Runge-Kutta command

is designed for systems of equations. For a scalar equation of the form $y' = r(x, y)$, the command is

$$\boxed{\texttt{RK([r(x,y)],[x,y],[x0,y0],h,n)}} \; .$$

Therefore, for this problem the Runge-Kutta method is set up with the `Author` command and by entering

$$\boxed{\texttt{RK([(y\^{}2 sin(2x))/x\^{}2],[x,y],[1,2],0.15,50)}} \; .$$

Execute the `approX` command to obtain a set of solution values. The command and its truncated result are as follows:

17: RK $\left[\left[\dfrac{y^2 \; \text{SIN} \; (2 \; x)}{x^2}\right], \; [x, \; y], \; [1, \; 2], \; 0.15, \; 50\right]$

18: [[1, 2], [1.15, 2.55904], [1.3, 3.06231], [1.45, 3.37962], [1.6, 3.45205],

Execute `Plot Plot` to see these values produced with the Runge-Kutta method plotted on the same axes as the values determined using the Euler method.

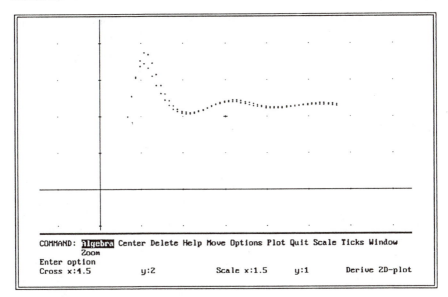

```
COMMAND: algebra Center Delete Help Move Options Plot Quit Scale Ticks Window
         Zoom
Enter option
Cross x:4.5          y:2              Scale x:1.5    y:1         Derive 2D-plot
```

The points are pretty close, although there is more difference toward the beginning (left side) of the interval.

What happens if we reduce the step size to 0.1 and compute for 75 steps? The result of plotting these 75 points on the same axes as the other two plots is as shown.

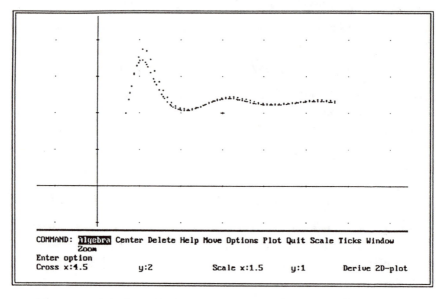

These numerical methods are useful when exact solutions cannot be determined. The EULER function is faster, but is subject to more error than the RK function. The RK function can also be used for systems of differential equations.

Example 4.3: Taylor Polynomial

... over the entrances to the gates of the temple of science are written the words: Ye must have faith.

—Max Planck [1932]

Subject: Taylor Polynomial Approximations

References: Sections 1.7, 1.10, 1.11, and 1.13

Problem: Find the Taylor polynomial approximations of degrees 3, 4, 7, and 9 for $\sin x$ about $x = 0$. Then find approximate solutions to the nonlinear differential equation

$$y' = \sin(xy),$$

with $y(0) = 1$, using 3rd-, 4th-, 7th-, and 9th-degree Taylor polynomials.

Solution: The Taylor polynomial approximation for a function can be obtained using Derive in two ways. The first way is to use the `Calculus Taylor` menu command and answer the queries for the variable, degree, and point. The second way is to use the in-line `TAYLOR` function that requires this information as arguments.

In order to obtain numerical coefficients for the polynomial, place Derive in the `Approximate` mode using the `Options Precision` command. Using the second method for the third-degree polynomial approximation, `Author`

$$\boxed{\texttt{TAYLOR(sinx,x,0,3)}}\ .$$

Then issue `Simplify` to obtain

```
1:   TAYLOR (SIN (x), x, 0, 3)

                      3
2:   x - 0.166666 x
```

Using the first method for the fourth-degree polynomial approximation, we `Author` $\boxed{\texttt{sinx}}$. Issue the `Calculus Taylor` command. Respond to the input queries with `Variable:` x, `Degree:` 4, and `Point:` 0. Execute the `Simplify` command to get

```
3:    SIN (x)

4:    TAYLOR (SIN (x), x, 0, 4)

                          3
5:    x - 0.166666 x
```

We see that the 3rd- and 4th-degree polynomial approximations are the same because the coefficient of the x^4 is evaluated as 0. Produce the 7th- and 9th-degree polynomials using either of the two methods. The results of these operations are as follows:

```
6:    TAYLOR (SIN (x), x, 0, 7)

                    -4  7                 5              3
7:    - 1.98412 10    x  + 0.00833333 x  - 0.166666 x  + x

8:    TAYLOR (SIN (x), x, 0, 9)

                -6  9          -4  7                 5              3
9:    2.75573 10    x  - 1.98412 10    x  + 0.00833333 x  - 0.166666 x  + x
```

Let's plot these polynomials (degrees 3, 7, and 9) along with the function $\sin x$ to compare their accuracy. Assemble the four functions into a vector with appropriate components by *Authoring* $\boxed{\texttt{[\#3,\#5,\#7,\#9]}}$. Then issue the `Plot Overlay Plot` command. The graphs using the default plotting parameters are as follows:

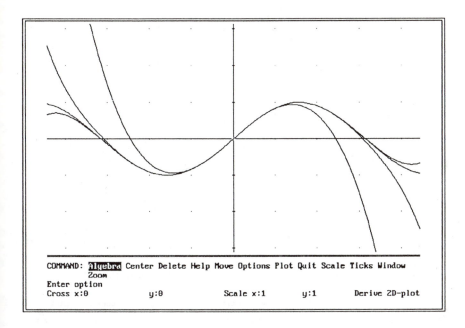

```
COMMAND: Algebra Center Delete Help Move Options Plot Quit Scale Ticks Window
         Zoom
Enter option
Cross x:0              y:0              Scale x:1      y:1       Derive 2D-plot
```

All of the polynomials seem to approximate $\sin x$ to within the plotting accuracy of the computer's graphics resolution in the interval $-1.25 < x < 1.25$, and the higher-degree polynomials seem to improve the accuracy outside this interval. To get a more global view of the functions, execute the Scale command and set the x-scale to 3. The larger-scaled plot is as follows:

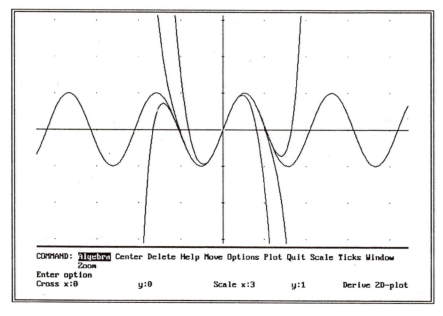

An approach to approximating solutions to nonlinear differential equations is to expand the functions into Taylor polynomials accurate in some neighborhood of the point of expansion. Derive has a command to find the nth-degree truncated Taylor-series solution in the utility file ODE_APPR. Once the file ODE_APPR is loaded into the work area with the **Transfer Load Utility** command, the command to approximate the solution to the given differential equation,

$$y' = \sin(xy),$$

with $y(0) = 1$, with a 3rd-degree polynomial is

$$\boxed{\texttt{TAY\_ODE1(sin(xy),x,y,0,1,3)}}\;.$$

Author and Simplify this operation to obtain the following result:

```
25:  TAY_ODE1 (SIN (x y), x, y, 0, 1, 3)

            2
26:  0.5 x  + 1
```

The higher-degree approximations are determined by a similar procedure. Since only the last argument in the command needs to be changed, the highlight and **F3** key can be helpful in reducing the retyping of the entire command. The higher-degree approximations take considerable time. The results of the three operations (degrees 4, 7, and 9) are as follows:

```
27:  TAY_ODE1 (SIN (x y), x, y, 0, 1, 4)

               4        2
28:  0.0833333 x  + 0.5 x  + 1

29:  TAY_ODE1 (SIN (x y), x, y, 0, 1, 7)

                6              4        2
30:  - 0.0263888 x  + 0.0833333 x  + 0.5 x  + 1

31:  TAY_ODE1 (SIN (x y), x, y, 0, 1, 9)

                8            6              4        2
32:  - 0.0215525 x  - 0.0263888 x  + 0.0833333 x  + 0.5 x  + 1
```

Once again, the easiest way to plot these functions all at once is to place them in a vector as components through Author of the input

[#26,#28,#30,#32] .

Then go to the Plot window and Scale the x-scale to 3 and the y-scale to 10. Now execute Plot. The plots of these functions reveal symmetry with respect to both axes and provide visualization of the solution behavior near the initial value $x = 0$.

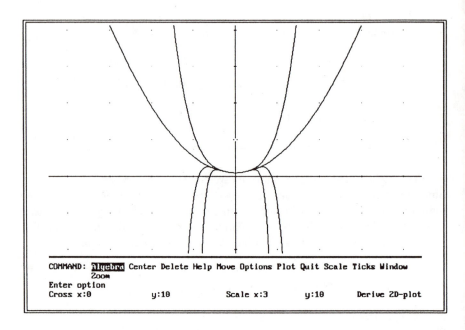

```
COMMAND: Algebra Center Delete Help Move Options Plot Quit Scale Ticks Window
         Zoom
Enter option
Cross x:0              y:10           Scale x:3       y:10       Derive 2D-plot
```

However, the solution trajectories are quite different away from the start point, which shows the poor nature of the approximations for $|x| > 1.2$. This technique is only good for analyzing local behavior of the solution.

Exercise 4.4: Taylor Series

The divergent series are the invention of the devil.

—Niels Henrik Abel [1826]

Subject: Approximating Functions with Taylor Series

Purpose: To use Derive to find the Taylor polynomial approximations for several functions.

References: Sections 1.7, 1.10, and 4.3

Given: The following functions, $f_i(t)$, $i = 1, 2, \ldots, 7$ and values of t:

$$
\begin{aligned}
f_1(t) &= t^2 e^t, & t &= 0, \\
f_2(t) &= \cos(t^2), & t &= 0, \\
f_3(t) &= \sec t, & t &= \pi/3, \\
f_4(t) &= t e^t, & t &= -2, \\
f_5(t) &= \ln(2 + t^2), & t &= 0, \\
f_6(t) &= 1/t, & t &= 3, \\
f_7(t) &= 2^t, & t &= 0.
\end{aligned}
$$

Exercises:

1. Use the function TAYLOR to find the first three terms of the Taylor series about the given value for t for these 7 functions.

2. Plot each of the 7 functions and its 3-term Taylor approximation in the interval $-4 < x < 4$. What are the periods of these functions? Which functions are even? Which are odd?

3. Find the first 6 terms for each of these 7 functions. Plot the function, the 3-term approximation, and the 6-term approximation for each of the functions on the interval $-4 < x < 4$. For each, does the approximation improve with more terms?

4. Approximate the value of $\int_{-1}^{1} f_7(t)\, dt$ using both the 3-term and 6-term approximations for f_7. Find a more accurate value for this integral and compare the two approximations with that value.

Example 4.5: Difference Equation

I must study politics and war that my sons may have liberty to study mathematics and philosophy.

—John Adams [1780]

Subject: Using Difference Equations for a Model in Personal Finance

References: Sections 1.6 and 1.14

Introduction: This problem is an exploration of several different scenarios involving difference-equation models for finance. Difference equations are the discrete mathematics analog of the differential equations in continuous mathematics. The scenarios involve first-order difference equations as models for accumulating and withdrawing money from savings accounts.

If $D(n)$ is the number of dollars in a savings account after n compounding periods, the account collects interest at a $100i$ percent annual rate compounded m times a year, and a dollars are added to the account at the end of each compounding period, then a difference-equation model for this situation is

$$D(n + 1) = (1 + i/m)D(n) + a.$$

If a is negative, then money is withdrawn from the account. An account with continuous compounding is modeled by a differential equation.

The Derive utility file RECUREQN.MTH has several command functions that solve for the analytic solution of some of the common first- and second-order difference equations. Other names commonly used for difference equations are discrete dynamical systems and recurrence relations. The Derive commands ITERATE and ITERATES can perform the iteration of difference equations to determine specific terms in the solution. ITERATES produces all the iterations performed, whereas ITERATE only outputs the last iteration performed. The following table provides a short description of some of the commands in this utility file:

Command	Function
LIN1_DIFFERENCE (p,q,n,n0,y0)	Solves the first-order nonhomogeneous linear difference equation $y(n+1) = p(n)y(n) + q(n)$ with $y(n_0) = y_0$.
GEOMETRIC1 (p,q,k,n,n0,y0)	Solves the linear difference equation $y(kn) = p(n)y(n) + q(n)$, $y(n_0) = y_0$.
RECURRENCE (r,n,y,n0,y0,k)	Produces a matrix of k iterates of the equation $y(n+1) = r(n, y(n))$ with $y(n_0) = y_0$.
LIN2_CCF(p,q,r,n)	Solves the second-order, constant-coefficient difference equation $y(n+2) + py(n+1) + qy(n) = r(n)$.
LIN2_CCF_BV (p,q,r,n,n0,y0,n2,y2)	Solves the 2nd-order difference equation with boundary conditions $y(n_0) = y_0$ and $y(n_2) = y_2$.

Commands for solving difference equations in
utility file RECUREQN.MTH.

Therefore, the command to solve the above first-order equation with condition $D(k) = D_k$ is LIN1_DIFFERENCE(1+i/m,a,n,k,Dk).

Problem A: Now, let's tackle the first scenario. If there are two banks offering competing savings plans where one plan has a 5.125% interest compounded yearly and the other plan has a 5% interest compounded daily, which bank has the better deal for the customer?

Solution to A: The model for the daily compounding in this scenario is

$$D(n+1) = (1 + 0.05/365)D(n),$$

with $D(0) = c$, where c represents the amount of money initially deposited in the bank. Once the utility file is loaded into the work area with commands Transfer Load Utility and filename $\boxed{\text{RECUREQN}}$, the difference equation is entered into the work area with the Author command and typing

$$\boxed{\text{LIN1\_DIFFERENCE(1+0.05/365,0,n,0,c)}}.$$

Simplify to get the analytic solution to the difference equation.

$$34:\quad \text{LIN1\_DIFFERENCE}\left[1 + \frac{0.05}{365},\ 0,\ n,\ 0,\ c\right]$$

$$35:\quad c\left[\frac{7301}{7300}\right]^n$$

This result is messy and could be cleaned up by executing the Manage Exponential Collect command and then the Simplify command. In this case, we don't care since we just want to know the balance after 1 year or 365 days, so Manage Substitute for n the value 365. Just press **Enter** when the value for the variable c is asked for. It is probably best to change the arithmetic precision to Approximate through the Options Precision command before the next operation. This will save the computation of a fraction with hundreds of digits in the numerator and denominator. Then, Simplify this expression to obtain

$$36:\quad c\left[\frac{7301}{7300}\right]^{365}$$

$$37:\quad 1.05126\ c$$

From this result, we see that the effective interest rate is higher (5.126%) for the bank with the daily compounding and a lower annual rate.

Problem B: The next scenario involves developing a lifetime savings/retirement plan. You want to retire in 35 years, and in that time save enough so you can withdraw the equivalent of $30,000 in current money for the next 25 years of retirement. What is the equivalent of $30,000 in 35 years?

Solution to B: If 6% annual inflation is assumed, the difference-equation model to determine the equivalent of $30,000 in 35 years is

$$D(n+1) = (1 + 0.06/1)D(n),$$

with $D(0) = 30{,}000$. To solve this, Author and Simplify the command

LIN1_DIFFERENCE(1+0.06,0,n,0,30000) .

Then `Manage Substitute` 35 for n and `approX`. Derive responds with the following output:

```
38:   LIN1_DIFFERENCE (1 + 0.06, 0, n, 0, 30000)

            0.0582689 n
39:   30000 ê

            0.0582689 35
40:   30000 ê

            5
41:   2.30582 10
```

Wow! You'll need over $230,000 a year for each of your 25 years (or a total of $5,764,500) to retire.

Problem C: Can you save enough to have that much in 35 years? How much do you need to save?

Solution to C: Assume you will invest all your savings when you retire at 10% interest compounded annually and withdraw $230,000 per year for 25 years from this account. At that time, you will have exhausted your savings and the balance will be $0. The difference equation model for this plan is

$$D(n + 1) = (1 + 0.1/1)D(n) - 230000,$$

with $D(25) = 0$. Once again, `Author` and `Simplify` the expression

LIN1_DIFFERENCE(1+0.1,-230000,n,25,0) .

The resulting Derive display is shown.

```
42:  LIN1_DIFFERENCE (1 + 0.1, -230000, n, 25, 0)

               6              5  0.0953101 n
43:  2.3 10  - 2.1228 10   ê
```

What is $D(0)$? **Manage Substitute** 0 for n and **Simplify** to get

```
               6              5  0.0953101 0
44:  2.3 10  - 2.1228 10   ê

               6
45:  2.08772 10
```

Surprise! You need to save $2,000,000 during the next 35 years. At least this is considerably better than $5,764,500.

Problem D: How much do you have to save each month?

Solution to D: Assume you will invest in a savings account that pays 8% interest compounded monthly and will deposit p dollars every month for 420 months (35 years). A practical model for this is

$$D(n + 1) = (1 + 0.08/12)D(n) + b,$$

with $D(420) = 2{,}000{,}000$. **Author** this into Derive with the command

LIN1_DIFFERENCE(1 + 0.08/12,b,n,420,2000000) .

Simplify this expression to get the solution. **Manage Substitute** 0 for n. Return to the **Exact** mode using **Options Precision** and **soLve** for b. Then, approX this result. The expressions in the work area to do these steps are shown.

$$46: \quad \text{LIN1\_DIFFERENCE} \left[1 + \frac{0.08}{12}, b, n, 420, 2000000\right]$$

$$47: \quad 3.06891 \ (3 \ b + 40000) \ \hat{e}^{0.00664451 \ n} - 150 \ b$$

$$48: \quad 3.06891 \ (3 \ b + 40000) \ \hat{e}^{0.00664451 \ 0} - 150 \ b$$

$$49: \quad b = \frac{476020000}{545961}$$

$$50: \quad b = 871.893$$

Now you know that you need to save about \$870 per month for the next 35 years to retire on the equivalent of \$30,000 per year for the following 25 years.

Exercise 4.6: Second-Order Difference Equation

Modern science ... is an education fitted to promote sound citizenship.

—Karl Pearson [1892]

Subject: Second-Order Difference Equation for the National Economy

Purpose: To solve and analyze a difference equation model for the national economy.

References: Sections 1.10 and 4.5

Given: The national income during time period n is denoted by $I(n)$ and can be modeled as the sum of incomes of consumer expenditures, private investment, and government expenditures. While these components exist continuously, they are only known and predicted at discrete periods of time. By making assumptions on the behavior of these components over a time period, the following difference equation can be used to model the national income:

$$I(n+2) = (1+a)cI(n+1) - acI(n).$$

In this model a is the constant of adjustment, and c is the marginal propensity to consume. Valid values for these two constants are $a > 0$ and $0 < c < 1$.

Exercises:

1. Use commands from the utility file RECUREQN.MTH to solve this model for $a = 1.1$ and $c = 0.3$.

2. If $I(0) = 230$ and $I(1) = 248$, find the solution and plot it for $0 < n < 12$.

3. If $I(0) = 230$ and $I(1) = 270$, find the solution and plot it for $0 < n < 12$ on the same graph as the solution found in #2.

4. What is the value of $I(200)$ from the solution obtained in #3?

5

Second- and Higher-Order Equations

The emphasis should be on concepts, understanding, and applications. The CASs can handle the routine computations while the student directs the analysis.

—Donald Small and John Hosack [1990]

Sample problems in second- and higher-order differential equations have been worked out to show the power and versatility of Derive as a problem-solving tool. By carefully reading these problems and working along with Derive, you should get a better feel for the subjects of differential equations and mathematical problem solving. Some of these examples also show the limitations of the computer software and therefore help to determine when and when not to use Derive.

Some of the examples involve models of applications, while others are posed in a mathematical context. The problems are similar to those typically found in undergraduate applied differential equations textbooks.

These exercises are to be done by you to direct your learning of new knowledge about differential equations, Derive, and mathematics in general. Some of the exercises in this chapter lead you to explore the mathematics and software and to discover new results on your own. As you solve the problems from these exercises, ask yourself "what if" questions and answer them.

Example 5.1: Projectile Motion

Come one, come all! this rock shall fly
From its firm base as soon as I.

—Sir Walter Scott [1810]

Subject: Projectile Motion in Two Dimensions

References: Sections 1.6, 1.8, and 1.10

Problem: A projectile is fired over horizontal ground at an initial speed of v meters/second at an angle of elevation a. a is between 0 and $\pi/2$ radians. Where will the projectile be t seconds after the firing? Assume the only force acting on the projectile is gravity, which produces a downward acceleration of 9.8 meters/sec^2.

Solution: First, we let $x(t)$ represent the horizontal displacement and $y(t)$ represent the vertical displacement. We make the firing point the origin (0,0). To start our calculations, let's use a special case of $a = \pi/2$ radians (90°). This is the case of the projectile being fired straight up, and the problem reduces to one space dimension y.

Start with acceleration $y'' = -9.8$. Integrate this to find the velocity $y'(t) = -9.8t + c_1$ (you shouldn't need to use Derive for this easy integration). Since $y'(0) = v$, $c_1 = v$ and $y'(t) = -9.8t + v$. The vertical displacement y is found by another integration. This time, use Derive and Author $\boxed{\texttt{INT(-9.8t+v,t)}}$. Simplify to obtain the following result:

$$1: \quad \int (-9.8\,t + v)\,dt$$

$$2: \quad t\,v - \frac{49\,t^2}{10}$$

Remember, Derive does not provide a constant of integration, so $y = -4.9t^2 + vt + c_2$. Since the firing takes place at the origin, $y(0) = 0$, which makes $c_2 = 0$. Therefore, the height of the projectile at time t is simply $y(t) = -4.9t^2 + vt$.

Now, we will use vector equations to solve the problem for any angle a between 0 and $\pi/2$ radians. Start with the acceleration due to gravity in vector form $[x'', y''] = [0, -9.8]$. Integrate to find the velocity vector $[c_1, -9.8t + c_2]$. Once again, this is done without Derive. The initial velocity vector is $[v\cos a, v\sin a]$, which makes $c_1 = v\cos a$ and $c_2 = v\sin a$. Therefore, the velocity vector $[x', y']$ is $[v\cos a, -9.8t + v\sin a]$. To find the displacement vector $[x, y]$, we need to integrate the velocity vector. **Author**

```
INT([v cosa, -9.8t+v sina],t)
```

and **Simplify**. The result is

```
3:   ∫ [v COS (a), - 9.8 t + v SIN (a)] dt

                                      2
                                  49 t
4:   [t v COS (a), t v SIN (a) - ─────]
                                   10
```

Add the constants of integration, b and c, to the respective components of the vector. This can be done through the **Author** command by either retyping the expression with the appropriate constants or by using the **F3** key to bring the highlighted expression into the author line of the menu and editing it. In order to edit an expression in the author line, we use the **Ctrl-S** keys to move the cursor to the left and the **Ctrl-D** keys to move the cursor to the right. The **Ins** key can be used to toggle the insert mode on and off. Press the **Enter** key when the expression is completed. The work area now contains the following expressions:

$$5: \quad \left[t \ v \ COS \ (a) \ + \ b, \ t \ v \ SIN \ (a) \ - \ \frac{49 \ t^2}{10} \ + \ c \right]$$

First we use Manage Substitute to replace t with 0, and Simplify. Then, since the initial displacements were (0,0), we issue the soLve command. The result is the following trivial values for these two constants:

$$6: \quad \left[0 \ v \ COS \ (a) \ + \ b, \ 0 \ v \ SIN \ (a) \ - \ \frac{49 \ 0^2}{10} \ + \ c \right]$$

$$7: \quad [b, \ c]$$

$$8: \quad [b = 0, \ c = 0]$$

Therefore, the displacement vector in meters of a projectile fired at an angle a is $[(v \cos a)t, -4.9t^2 + (v \sin a)t]$. Let's plot this parametric vector for several values of a for $0 < t < 10$. We choose $v = 100$ and $a = \pi/8, \pi/4, 3\pi/8$, and $9\pi/20$. We have to plot these one at a time, going back and forth from the algebra window to the plot window.

First, we Author

$$[(100\cos a)t, \ -4.9t\verb|^|2+(100\sin a)t] \ .$$

Then select Manage Substitute to replace a with pi/8 and just press Enter when queried for a value of t. Then select the Plot menu and Overlay location. Before plotting the function, we need to rescale. Select Scale and set x-scale = 200 and y-scale = 200. Issue Plot. Derive realizes the highlighted function is in parametric form, so it produces a submenu to set up several parametric plotting values. Since we want $0 < t < 10$, set Min at 0 and Max at 10 and press **Enter**. The resulting plot is as follows:

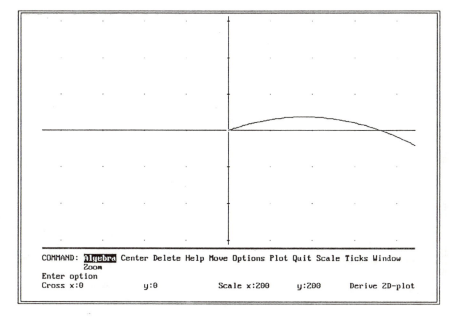

In order to produce each of the other three plots on the same graph, select **Algebra** and repeat the entire process using the remaining three different values of a. Each time the **Plot** menu is selected, all the previous plots will be produced before the new plot is drawn. This is a time-consuming process, and the slowness of plotting is one of the limitations of Derive. The resulting plot screen is as follows:

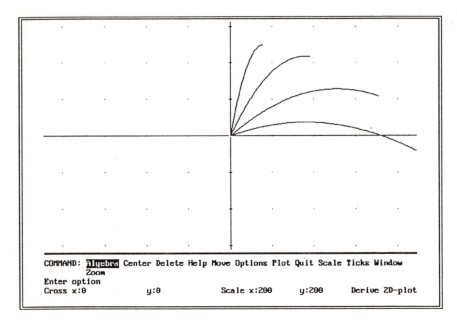

```
COMMAND: Algebra Center Delete Help Move Options Plot Quit Scale Ticks Window
         Zoom
Enter option
Cross x:0          y:0               Scale x:200     y:200      Derive 2D-plot
```

The resulting algebra work area is as follows:

11: $\left[100 \cos \left[\dfrac{\pi}{8} \right] t, \ -4.9 \ t^2 + 100 \sin \left[\dfrac{\pi}{8} \right] t \right]$

12: $\left[100 \cos \left[\dfrac{\pi}{4} \right] t, \ -4.9 \ t^2 + 100 \sin \left[\dfrac{\pi}{4} \right] t \right]$

13: $\left[100 \cos \left[\dfrac{3\pi}{8} \right] t, \ -4.9 \ t^2 + 100 \sin \left[\dfrac{3\pi}{8} \right] t \right]$

14: $\left[100 \cos \left[\dfrac{9\pi}{20} \right] t, \ -4.9 \ t^2 + 100 \sin \left[\dfrac{9\pi}{20} \right] t \right]$

Example 5.2: Undetermined Coefficients

If you do not rest on the good foundation of nature, you will labor with little honor and less profit.

—Leonardo da Vinci

Subject: Solving a Complete, Third-Order, Constant-Coefficient, Initial-Value Problem with Undetermined Coefficients

References: Sections 1.7, 1.8, and 1.10

Problem: Find the solution to the following initial-value problem using undetermined coefficients:

$$y''' + y'' + y' + y = e^{-x} + kx,$$
$$y(1) = 1, \ y'(1) = 1/4, \ y''(1) = -1/2, \ -8 \le k \le 8.$$

Analyze the solution to determine the value of k that maximizes the solution for $x > 1$, and plot this solution.

Solution: The cubic auxiliary equation for this third-order differential equation is $m^3 + m^2 + m + 1 = 0$. Its roots are obtained by inputting this equation using the **Author** command and solving the equation using the **soLve** command. The output from these steps is

```
            3    2
    1:   m  + m  + m + 1 = 0

    2:   m = -1

    3:   m = -î

    4:   m = î
```

Therefore, the complementary solution of the homogeneous equation is $y_c = c_1 e^{-x} + c_2 \cos x + c_3 \sin x$.

By using the forms of the parts of the forcing function $e^{-x} + kx$ and this complementary solution, the form of a particular solution y_p can be written by using the method of undetermined coefficients as $y_p = Axe^{-x} + Bx + C$. Let $L[y] = y''' + y'' + y' + y$. Now, in order to determine coefficients A, B, and C, $L[y_p]$ is evaluated and equated to the forcing function. To do this with Derive, the operator $L[y]$ is defined using the in-line command for differentiation, DIF (see Section 1.7). This entry is done with the **Author** command and by input of the string

$$\boxed{\texttt{L(y):=DIF(y,x,3)+DIF(y,x,2)+DIF(y,x)+y}}$$.

The output display is

$$5: \quad L(y) := \left[\frac{d}{dx}\right]^3 y + \left[\frac{d}{dx}\right]^2 y + \frac{d}{dx} y + y$$

Now, this differential operator can be evaluated for y_p through **Author** of the expression

```
AUTHOR expression: l(axê^-x+bx+c)_

Enter expression
User                                    Free:100%              Derive Algebra
```

The working area display of this command and the result obtained by executing **Simplify** are shown.

$$6: \quad L\,(a\,x\,\hat{e}^{-x} + b\,x + c)$$

$$7: \quad 2\,a\,\hat{e}^{-x} + b\,x + b + c$$

Equating like coefficients results in three simultaneous equations: $2A = 1$, $B = k$, and $B + C = 0$. Their solution is obviously $A = 1/2$, $B = k$, and $C = -k$. If a more complicated system of equations for A, B, and C had to be solved, Derive could have been used to solve the system of equations for these unknowns (see Section 1.8).

The general solution, y_g, is the sum $y_c + y_p$, and can be authored as

$$\boxed{\texttt{y(x):=aAlt-e\^{}-x+b cosx+c sinx+0.5xAlt-e\^{}-x+kx-k}}$$.

This operator displays as

```
10:  Y (x) := a ê^(-x)  + b COS (x) + c SIN (x) + 0.5 x ê^(-x)  + k x - k
```

Because of Derive's restriction on subscripts, a, b, and c have been used in place of c_1, c_2, and c_3. In order to check y_g in the original equation, simply Author and Simplify the expression $\boxed{L(y(x))}$. The output is

```
11:   L (Y (x))


          -x
12:   ê     + k x
```

which shows that the operations on y_g contained in the left side of the equation yield an expression that is equal to the forcing term on the right side, and therefore y_g satisfies the equation.

Even before the constants a, b, and c are determined, the value of k that maximizes the solution for $x > 1$ can be determined. The parameter k appears in the solution in the form $kx - k$. For $x > 1$ and $-8 \leq k \leq 8$, $k = 8$ maximizes this expression and, therefore, the solution. Now, $y(x)$ must be redefined with 8 in place of k by executing Author and entering

```
AUTHOR expression: Y(x):=aê^(-x)+bCOS(x) + c SIN(x) + 8.5x ê^(-x) + 8 x - 8_

Enter expression
User                                    Free:100%           Derive Algebra
```

Implementing the initial conditions to find values for a, b, and c can also be done easily now that $y(x)$ is a defined function. Simply Author the command $\boxed{\texttt{y(1)=1}}$ and select $\texttt{Simplify}$ to obtain the first condition. The output is

$$15: \quad Y\ (1)\ =\ 1$$

$$16: \quad \left[a\ +\ \frac{1}{2}\right]\ \hat{e}^{-1}\ +\ b\ \text{COS}\ (1)\ +\ c\ \text{SIN}\ (1)\ =\ 1$$

The second condition, $y'(1) = 1/4$, is input by the in-line command $\boxed{\texttt{DIF(y(x),x) = 1/4}}$. $\texttt{Simplify}$ results in the following display:

$$17: \quad \frac{d}{dx}\ Y\ (x)\ =\ \frac{1}{4}$$

$$18: \quad -\ \frac{(x\ +\ 2\ a\ -\ 1)\ \hat{e}^{-x}}{2}\ +\ c\ \text{COS}\ (x)\ -\ b\ \text{SIN}\ (x)\ +\ 8\ =\ \frac{1}{4}$$

$\texttt{Manage Substitute}$ the value 1 for x to get

$$19: \quad -\ \frac{(1\ +\ 2\ a\ -\ 1)\ \hat{e}^{-1}}{2}\ +\ c\ \text{COS}\ (1)\ -\ b\ \text{SIN}\ (1)\ +\ 8\ =\ \frac{1}{4}$$

The third and last condition, $y''(1) = -1/2$, is implemented in a similar manner. The in-line command is

$$\boxed{\texttt{DIF(y(x),x,2)\ =\ -1/2}}\ .$$

Once again, 1 is substituted for x using the **Manage Substitute** command. The resulting equations are

$$20: \quad \left[\frac{d}{dx}\right]^2 Y(x) = -\frac{1}{2}$$

$$21: \quad \frac{(x + 2a - 2)\,\hat{e}^{-x}}{2} - b\,\text{COS}(x) - c\,\text{SIN}(x) = -\frac{1}{2}$$

$$22: \quad \frac{(1 + 2a - 2)\,\hat{e}^{-1}}{2} - b\,\text{COS}(1) - c\,\text{SIN}(1) = -\frac{1}{2}$$

The three equations for the initial conditions can be solved by placing them as components into a vector. One way to do this is to **Author** the expression [#m,#n,#p], where m, n, and p are the Derive statement numbers of the equations. For the screen displays shown above, the expression would be [#16,#19,#22]. Since we will be satisfied with decimal approximations for the coefficients in the final solution, **approX** this expression before executing the **soLve** command. The output of these steps is

$$23: \quad \left[\left[a + \frac{1}{2}\right]^{-1}\hat{e}^{-1} + b\,\text{COS}(1) + c\,\text{SIN}(1) = 1, \; -\frac{(1 + 2a - 1)\,\hat{e}^{-1}}{2} + c\,\text{COS} ($$

$$24: \quad [2.82181 \cdot 10^{-13} \;\; (1.30369 \cdot 10^{12} \; a + 1.91473 \cdot 10^{12} \; b + 2.98202 \cdot 10^{12} \; c + 6.51849$$

$$25: \quad \left[a = \frac{1003533115}{1476716814}, \; b = \frac{40536167418580997858411456729073}{61261794762839477287724269500}, \; c = -\frac{2842417686}{7948717870}\right]$$

$$26: \quad [a = 0.679570, \; b = 6.61687, \; c = -3.57594]$$

Now these values for the constants can be substituted into y_g to produce the solution

$$y = 0.679575e^{-x} + 6.61688\cos x - 3.57594\sin x + 0.5xe^{-x} + 8x - 8.$$

To plot this solution, **Author** it into the work area. The various parts can be pulled into the work area using the highlight and the **F3** key. Before issuing the **Plot** command, ensure the **Mode** in the **Options Display** menu is

set to Graphics. Execute the menu command Plot and select the Overlay
location. This sets up a 2-D plotting screen. Change the scale so the tick
marks in the x- and y-directions are more than the default of 1 unit. Since
there is an exponential function of x in the solution, it may be wise to have
a larger y-scale. Execute the Scale command and try an x-scale of 2 and a
y-scale of 25. Then issue Plot to get the following screen:

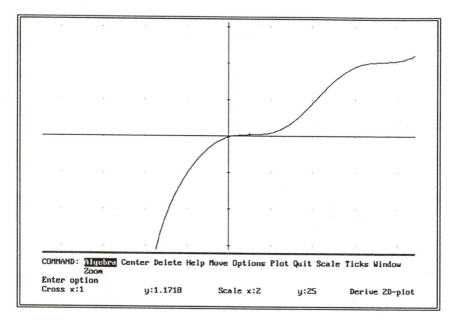

Using the Zoom and movable cross with the Center command enables the plot
range to be adjusted to any region of interest. Don't be impatient; it takes a
minute or two for the plot of this function to be completed.

Exercise 5.3: Undetermined Coefficients for Second-Order Equations

It is a great nuisance that knowledge can only be acquired by hard work.

—W. Somerset Maugham

Subject: Solving a Nonhomogeneous, Linear, Second-Order Differential Equation with Constant Coefficients Using Undetermined Coefficients

Purpose: To find the solution of an initial-value problem with parameters using the technique of undetermined coefficients and analyze the effect of varying the parameters.

References: Sections 1.7, 1.8, 1.10, and 5.2

Given:

$$y'' - 2y' + 10y = 3xe^{2x}\cos(3x), \tag{1}$$

with initial conditions $y(\pi/2) = 0$ and $y'(\pi/2) = a$.

Exercises:

1. Formulate the auxiliary equation and solve for its roots using Derive. Write the complementary solution, y_c, from these roots.

2. Using the method of undetermined coefficients, determine the form of a particular solution, y_p.

3. Let $L[y] = y'' - 2y' + 10y$. Using Derive, evaluate $L[y_p]$ and equate it to the right-hand side of the given nonhomogeneous equation (1).

4. Using Derive, solve for the undetermined coefficients by equating the like coefficients.

5. Write the general solution, y_g, from the functions y_c and y_p. Check your solution using Derive and the operator $L(y)$.

6. Substitute the initial conditions into y_g and solve for the arbitrary constants using Derive.

7. Plot the solutions to the initial-value problem for $-4 \leq x \leq 4$ and the parameter values of $a = -10$, 0, and 10.

8. Which of the three given values for a produces a maximum at $x = 0$?

9. What is the significance of the parameter in this problem? Relate this model with its parameter to a problem in engineering.

10. Ask yourself other questions related to this problem that you would like to investigate. Determine if Derive is a proper tool to use in this investigation by trying to answer your questions.

Example 5.4: Second-Order Differential Equations

Numerical precision is the very soul of science.

—Sir D'Arcy Wentworth Thompson [1917]

Subject: Solving a Second-Order, Constant-Coefficient, Nonhomogeneous Equation with a Forcing Function Using Utility File ODE2

References: Sections 1.10 and 1.13

Problem: Find the general solution to

$$y'' - 2y' + y = 12e^x - 4xe^x,$$

with $y(0) = 1$ and $y'(0) = 2$.

Solution: The operations for solving this kind of second-order, nonhomogeneous differential equation with constant coefficients are found in the utility file ODE2. Load this file with the **Transfer Load Utility** command.

In order to find the general solution of a second-order, linear, differential equation with initial conditions, the **DSOLVE2_IV** command is used. If the equation is in the form

$$y'' + p(x)y' + q(x)y = r(x),$$

with $y(x_0) = y_0$ and $y'(x_0) = v_0$, the form of the command is

DSOLVE2_IV(p,q,r,x,x0,y0,v0) .

Therefore, to solve the differential equation given in this problem, **Author** the command

DSOLVE2_IV(-2,1,12 **Alt-e^x-4xAlt-e^x**,x,0,1,2) .

Select Simplify to obtain the general solution as shown.

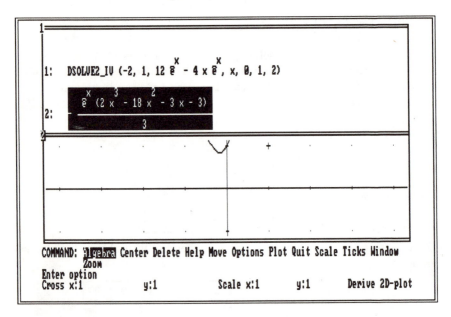

The plot of this solution with the default plotting parameters using the Plot Under Plot command produces

The scale needs to be adjusted to see the global nature of the solution. We try setting the Scale in the x-direction to 1 and the y-direction to 20000, using the Move command to move the cross to (2,5000), and executing the Center command to better center the interesting region of the plot on the screen. These plotting parameters produce a better graph that shows the exponential growth of the solution. This plot is shown below.

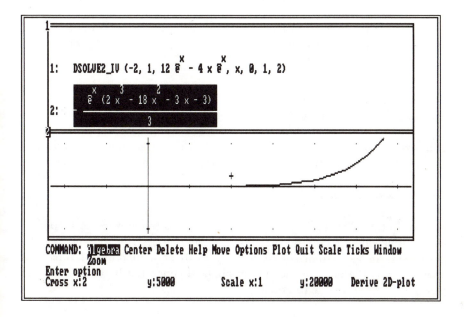

Exercise 5.5: Automobile Suspension System

> *Every body continues in its state of rest, or in uniform motion in a right line, unless it is compelled to change that state by forces impressed upon it.*

> —Isaac Newton

Subject: Vibrations in an Automobile Suspension System

Purpose: To model a physical mechanism with a second-order, nonhomogeneous differential equation, solve the equation, and analyze the behavior.

References: Sections 1.10 and 1.13

Given: In the following idealized drawing, an automobile wheel is supported by a spring and shock absorber. On the diagram, $y(t)$ is the vertical displacement (in feet) from equilibrium ($y = 0$) as a function of time t in seconds, m is the mass of the car (in pounds) supported by the wheel, and g is the acceleration due to gravity.

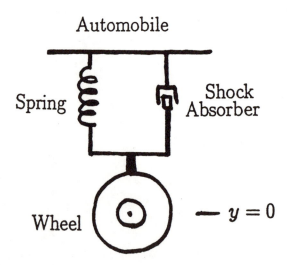

Assume viscous damping in the shock absorber (i.e., the frictional force is proportional to the velocity), Hooke's law for the restoring force of the spring, and a forcing function, $f(t)$, which represents the road conditions.

Exercises:

1. Write a second-order, constant-coefficient, nonhomogeneous differential equation for $y(t)$.

2. If a car weighing 3600 lbs is supported equally by four wheels with a spring constant of 200 lbs/ft, a nonnegative resistance coefficient of d, and a forcing function $f(t) = 200 \sin 6t$, write and solve the model for $y(t)$.

3. Assume further the case of no initial vertical displacement or velocity, and solve for $y(t)$.

4. Plot $y(t)$ for three values of the resistance coefficient of $d = 0$ lb-sec/ft, $d = 70$ lb-sec/ft, and $d = 140$ lb-sec/ft in the interval $0 < t < 4$ seconds.

5. Compare the three plots. Which case produces the largest vertical displacement in the interval? Which is the best shock absorber for this road condition?

6. Ask yourself other questions related to this problem that you would like to investigate. Determine if Derive is a proper tool to use in this investigation by trying to answer your questions.

Example 5.6: Safe Electrical Circuit

> *... the aim of exact science is to reduce the problems of nature to the determination of quantities by operations with numbers.*

> —James Maxwell [1856]

Subject: The Series Electrical Circuit

References: Sections 1.10 and 1.13

Problem: The basic, simple-loop electrical circuit consists of several parts: a time-dependent voltage source $E(t)$, an inductor L, a resistor R, and a capacitor C, arranged in series. Variations of this circuit containing more than one inductor, resistor, and capacitor are found throughout homes. In these simple circuits, L, R, and C represent the loop's total induction, resistance, and capacitance, respectively. The problem is to insure that the electrical current in any loop does not exceed the safe wire capacity that is protected by the limit of the fuse or circuit breaker.

Solution: Of course, we could take the experimental approach of plugging everything into the circuit and seeing if the fuse blows. However, we choose a more analytical approach. The best way to determine the current is to set up a differential model using Kirchoff's laws. By letting $q(t)$ be the time-dependent charge on the capacitor, the model is

$$L\frac{d^2q}{dt^2} + R\frac{dq}{dt} + \frac{1}{C}q = E(t),$$

where L, R, and C are assumed to be constant for this circuit.

The applied voltage for a typical US household alternating current may be modeled as

$$E(t) = 110\cos 2t \quad \text{volts.}$$

The Derive plot of this voltage is made with the **Author** command and by entering this expression $\boxed{\texttt{110 cos(2t)}}$ into the work area and using the **Plot Under** command. Good parameters for the **Scale** command are x-scale $= 5$ and y-scale $= 100$. **Move** the cross to $x = 20$ and $y = 0$, and execute the **Center** command. Don't forget to use the **Tab** key in order to move the cursor between different parts of the menu. After these parameters are set, the **Plot** command is issued to begin drawing the graph of the function. The resulting voltage plot is shown.

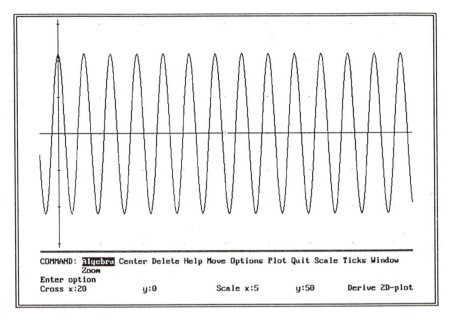

For a circuit with $L = 10$ henrys, $R = 50$ ohms, $C = 1/60$ farads, no initial charge on the capacitor $(q(0) = 0)$, no initial current $(q'(0) = 0)$, and $E(t)$ applied at $t = 0$ (q measured in volts, t measured in seconds), the equation to be solved is

$$10\frac{d^2q}{dt^2} + 50\frac{dq}{dt} + 60q = 110\cos 2t,$$

with $q(0) = 0$ and $q'(0) = 0$.

To solve this equation with Derive, first use the **Transfer Load Utility** command to obtain the utility file ODE2.MTH, which has commands available to solve second-order, constant-coefficient, nonhomogeneous equations in the form $y'' + ay' + by = r$. (Note: We have used a and b instead of the p and q used in the file ODE2 because q has already been used as the dependent variable in the equation.) Next, place the equation in this standard form

$$\frac{d^2q}{dt^2} + 5\frac{dq}{dt} + 6q = 11\cos 2t$$

to identify $a = 5$ and $b = 6$. Use the command

DSOLVE2_IV(5,6,11cos(2t),t,0,0,0) .

The resulting Derive display after executing `Simplify` is as follows:

```
2:   DSOLVE2_IV (5, 6, 11 COS (2 t), t, 0, 0, 0)

           - 2 t         - 3 t
       11 e          33 e        11 COS (2 t)   55 SIN (2 t)
3:   - --------- + --------- + ------------ + ------------
          4            13           52             52
```

The solution to the differential equation can now be written as

$$q(t) = -\frac{11}{4}e^{-2t} + \frac{33}{13}e^{-3t} + \frac{11\cos 2t}{52} + \frac{55\sin 2t}{52}.$$

Before plotting this solution, it should be checked. Derive can help do this. `Author` the operator for the equation by entering

```
AUTHOR expression: 10 DIF (q, t, 2) + 50 DIF (q, t) + 60 q - 110 COS (2 t)_

Enter expression
User                    D:2.9              Free:89%           Derive Algebra
```

Use the `Manage Substitute` command to substitute the expression for the solution into the equation for the variable q. To do this, use the highlight and **F3** key to put the equation in the input line of the menu, or use the # expression reference number. The display for this resulting command is truncated on the right in the following figure because of its length:

```
             d  2       11    - 2 t    33    - 3 t    11 COS (2 t)   55 SIN (2 t)
46:   10   [---]  [[- ----] e      + ---- e      + ------------ + ------------]
             dt        4             13                 52             52
```

Use the `Simplify` command to verify that the function does, in fact, evaluate to 0.

A check of the initial condition $q(0) = 0$ is made by substituting 0 for t and evaluating the expression. These steps can be done using the

Manage Substitute command and the Simplify command. The Derive results confirm that this condition is satisfied, as shown.

$$48: \quad \frac{-11\,e^{-2\,\theta}}{4} + \frac{33\,e^{-3\,\theta}}{13} + \frac{11\,\cos\,(2\,\theta)}{52} + \frac{55\,\sin\,(2\,\theta)}{52}$$

49:

Since the current $i(t)$ is equal to $q'(t)$, use Derive to get an expression for $i(t)$. One way to do this is to type the expression $\boxed{i \ = \ \text{DIF}(\#n,t)}$, where n is the statement number of the expression for $q(t)$. Or instead of using $\#n$, just highlight the expression for $q(t)$ and move it into the appropriate place in the work area with the **F3** key. The function displays as

$$48: \quad i = \frac{d}{dt}\left[\left[-\frac{11}{4}\right]e^{-2\,t} + \frac{33}{13}e^{-3\,t} + \frac{11\,\cos\,(2\,t)}{52} + \frac{55\,\sin\,(2\,t)}{52}\right]$$

Simplify this to obtain the following expression for $i(t)$:

$$49: \quad i = \frac{11\,e^{-2\,t}}{2} - \frac{99\,e^{-3\,t}}{13} + \frac{55\,\cos\,(2\,t)}{26} - \frac{11\,\sin\,(2\,t)}{26}$$

The easiest way to check that this function for the current does not exceed a certain amperage rating is probably to plot this function for $t > 0$ with a proper scale to see the global behavior. Highlight the expression for $i(t)$ and issue the **Plot** command. Use the **Delete All** command, if there are previous plots being drawn on the axes. Set x-scale $= 5$ and y-scale $= 4$ with the **Scale** menu command. Move the cross to $x = 20$ and $y = 0$, and execute the **Center** command. Plot the function to obtain the following plot screen:

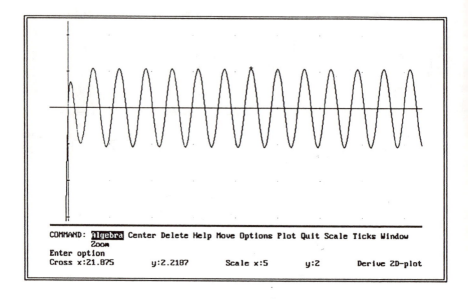

Move the cross using the direction arrows to determine that the maximum absolute current does not exceed a value of 2.3. Therefore, a 5-amp capacity (wiring and fuse) probably would be sufficient for this circuit.

While analyzing this solution, let's take note of the form of this long-term (steady-state) behavior. This behavior is more similar in form to the forcing function than the initial conditions. The initial conditions seem to affect only the short-term behavior. The graphing capabilities of Derive are very useful in making qualitative observations such as these. This can be explained by looking at the first two summands in the expression for $i(t)$. Both contain negative exponentials and therefore die off as t increases. Thus, the remaining trigonometric terms eventually dominate the solution.

Exercise 5.7: Electrical Circuit

The science of physics does not only give us an opportunity to solve problems, but it helps us also to discover the means of solving them, ...

—Jules Henri Poincaré

Subject: Electrical Circuit, Modeled as a Nonhomogeneous System and Solved Using Variation of Parameters

Purpose: To model an electrical circuit with a system of equations and solve the system using variation of parameters.

References: Sections 1.13 and 5.6

Given: The following electrical circuit diagram:

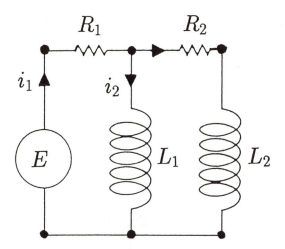

Exercises:

1. Use Kirchoff's first and second laws to write a first-order differential system in matrix notation for i_1 and i_2 in terms of R_1, R_2, L_1, L_2, and E, with
$$\vec{I} = \begin{bmatrix} i_1 \\ i_2 \end{bmatrix}.$$

2. If $R_2 = 3$ ohms, $L_1 = 1$ henry, and $L_2 = 1$ henry, find the eigenvalues and eigenvectors for this system in terms of R_1.

3. Write the complementary solution $\vec{I_c}$.

4. Is there a value of R_1 that produces a positive eigenvalue? Why or why not?

5. If $E = 50$ volts, find the particular solution $\vec{I_p}$ using variation of parameters (see Example 5.4).

6. Write the general solution $\vec{I_g}$.

7. If $i_1(0) = 0$ and $i_2(0) = 0$, write the solution $\vec{I}$.

8. What happens if $R_1 = 0.1$ ohms? if $R_1 = 4$ ohms? if $R_1 = 8$ ohms? Graph i_1 for these three cases on one set of axes $(0 \leq t \leq 10)$.

9. Ask yourself other questions related to this problem that you would like to investigate. Determine if Derive is a proper tool to use in this investigation by trying to answer your questions.

Exercise 5.8: Deflection of a Beam

Resources are the foundations upon which problem-solving performance is built.

—Alan Schoenfeld [1985]

Subject: Fourth-Order Differential-Equation Model for the Deflection of a Beam

Purpose: To solve a higher-order nonhomogeneous differential equation and to analyze the effects on beam deflection of varying parameters in the loading of the beam.

References: Sections 1.7 and 1.10

Given: The deflection of a beam carrying a load of $p(x)$ per unit length is modeled with the fourth-order equation

$$EI\frac{d^4w}{dx^4} = p(x),$$

where x is distance along the beam and E and I are positive constants. The physical characteristics of the beam are a length of 30, $E = 3.6$, $I = 2.71$, clamped on the left end ($x = 0$), and free on the right end ($x = 30$). Therefore, the four boundary conditions are $w(0) = 0$, $w'(0) = 0$, $w''(30) = 0$, and $w'''(30) = 0$.

Exercises:

1. Find and plot on the same axes $w(x)$ when $p(x) = (2x + a)/100$, for $a = 0, 10$, and 20. What is the total load on the beam for these values of a?

2. Find and plot on the same axes $w(x)$ when

$$p(x) = \begin{cases} 1, & \text{if } 0 < x < 10 \\ a, & \text{if } 10 < x < 20 \\ 1, & \text{if } 20 < x < 30 \end{cases},$$

for $a = 1, 5$, and 10. What is the total load on the beam for these values of a?

3. Find and plot on the same axes $w(x)$ when

$$p(x) = \begin{cases} 1, & \text{if } 0 < x < 20 \\ a, & \text{if } 20 < x < 30 \end{cases},$$

for $a = 1, 5$, and 10. What is the total load on the beam for these values of a?

4. If the beam is supported by an elastic foundation, the model becomes

$$EI\frac{d^4w}{dx^4} + kx = p(x),$$

where k is a positive constant. For a beam with $k = 4$ and the other physical values unchanged from the previous exercises, find and plot on the same axes $w(x)$ when $p(x) = (2x + a)/100$, for $a = 0, 10$, and 20. Compare these deflections with those found in #1. What is the effect of this elastic foundation on the deflection of the beam?

5. Ask yourself other questions related to this problem that you would like to investigate. Determine if Derive is a proper tool to use in this investigation by trying to answer your questions.

Example 5.9: Budget Growth

Men pass away but their deeds abide.

—Cauchy [1857]

Subject: Budget Growth and Conversion of a System of Equations to a Higher-Order Equation

References: Sections 1.7, 1.10, and 1.13

Problem: An organization's budget is divided into two categories, procurement and operations. Given the type of organization and by letting $e_1(t)$ be the procurement expenditure in thousands of dollars at time t (measured in years), $e_2(t)$ the operations expenditure in thousands of dollars at time t, and $g(t)$ the independent growth factor in thousands of dollars at time t, the time rate of change of the expenditures is modeled by the following system of differential equations:

$$\frac{de_1}{dt} = 0.1e_1 - 0.01e_2 + g(t)$$

and

$$\frac{de_2}{dt} = 0.3e_1 + 0.02e_2 + g(t).$$

If the current expenditures are $e_1(0) = 250$ and $e_2(0) = 114$ and if the growth function is a constant $g(t) = 2$, solve the differential model and analyze the future patterns in the organization's expenditures by graphing the solutions.

Solution: One way to solve a system of two linear differential equations is to convert it into a second-order equation with only one dependent variable by differentiating one of the equations and substituting the other into it. If this is done for this problem, we can use Derive's commands in ODE2.MTH to help solve the resulting second-order equation. One way to perform the manipulations to make the conversion is to use Derive.

Derive does not allow subscripted variables or function names, so we can't use $e_1(t)$ or $e_2(t)$ directly in Derive. However, in Word input mode, Derive does allow two or more characters in a name, like $E1$ and $E2$, so we'll use these names in place of the subscripted names used above. This mode is established with the Options Input Word menu command.

Start the conversion work by using the Declare Function command to make first $E1$ and then $E2$ functions of t. This command queries the user for the function name and its variables. The keystrokes to do this are

$$\boxed{\text{d f e1 Enter Enter t Enter}}$$

and

$$\boxed{\text{d f e2 Enter Enter t Enter}}.$$

After these declarations, the screen display will show

```
1:    E1 (t) :=

2:    E2 (t) :=
```

Next **Author** the two equations. The inputs to do this are

$$\boxed{\text{DIF(e1(t),t) = 0.1e1(t) - 0.01e2(t) + 2}}$$

and

$$\boxed{\text{DIF(e2(t),t) = 0.3e1(t) + 0.02e2(t) + 2}}.$$

The following image of the work area shows these operations:

```
        d
3:     ── E1 (t) = 0.1 E1 (t) - 0.01 E2 (t) + 2
        dt

        d
4:     ── E2 (t) = 0.3 E1 (t) + 0.02 E2 (t) + 2
        dt
```

The simplest Derive command to differentiate the entire equation for $\frac{de_1}{dt}$ is $\boxed{\text{DIF(\#n,t)}}$ where n is the number of the equation in the work area

(#3 in the work area for the screen above). Use Simplify to get the result shown.

$$5: \quad \frac{d}{dt} \left[\frac{d}{dt} \text{E1}(t) = 0.1 \text{ E1}(t) - 0.01 \text{ E2}(t) + 2 \right]$$

$$6: \quad \left[\frac{d}{dt} \right]^2 \text{E1}(t) = \frac{\frac{d}{dt} \text{E1}(t)}{10} - \frac{\frac{d}{dt} \text{E2}(t)}{100}$$

Now, substitute the right-hand side of the equation for $\frac{de_2}{dt}$ into the new second-order equation. Unfortunately, Derive does not allow substitution for a declared function (or its derivatives) through use of the Manage Substitute command. Therefore, the substitution must be done differently. The keystroking hints of Section 1.3 are very helpful. Execute the Author command and use the direction keys to move the highlight, the **F3** or **F4** key to copy highlighted text from the work area to the command line, and the **Ctrl-S** and **Ctrl-D** keystrokes to move left and right in the command line. If you have not done this kind of keystroking, now is the time to start. The alternative is retyping the entire input line. The result of this substitution is as follows:

$$8: \quad \left[\frac{d}{dt} \right]^2 \text{E1}(t) = \frac{\frac{d}{dt} \text{E1}(t)}{10} - \frac{0.3 \text{ E1}(t) + 0.02 \text{ E2}(t) + 2}{100}$$

This still leaves $e_2(t)$ in the equation, so this substitution procedure must be repeated for $e_2(t)$ in terms of $e_1(t)$ and $\frac{de_1}{dt}$ from the first equation, giving the following result. The line is truncated on the right by the screen width used by Derive.

$$10: \quad \left[\frac{d}{dt}\right]^2 E1(t) = \frac{\dfrac{d}{dt} E1(t)}{10} - \frac{0.3\ E1(t) + 0.02\left[-100\ \dfrac{d}{dt} E1(t) + 10\ E1(t\right.}{100}$$

Simplify this equation to obtain the second-order equation with constant coefficients for $e_1(t)$, as shown.

$$11: \quad \left[\frac{d}{dt}\right]^2 E1(t) = \frac{24\ \dfrac{d}{dt} E1(t) - E1(t) - 12}{200}$$

Now the commands in utility file ODE2.MTH can be used to solve this equation. Load this file into the work area with the **Transfer Load Utility** command. The equation has constant coefficients and can be expressed in the form used by file ODE2 as $y'' + py' + qy = r(x)$ with $E1$ substituted for y, $p = -24/200$, $q = 1/200$, and $r = -12/200$. Also, the independent variable for this problem is t instead of x. Therefore, the command to solve this equation is

$$\boxed{\text{DSOLVE2(-24/200,1/200,-12/200,t)}} \ .$$

The display of this command with the subsequent execution of the **Simplify** command gives the solution for e_1, as shown.

$$12: \quad \text{DSOLVE2}\left[-\frac{24}{200}, \frac{1}{200}, -\frac{12}{200}, t\right]$$

$$13: \quad \mathbf{e}^{3t/50}\left[c1 \cos\left[\frac{\sqrt{14}\ t}{100}\right] + c2 \sin\left[\frac{\sqrt{14}\ t}{100}\right]\right] - 12$$

The function for e_2 can be found from substitution of e_1 and $\frac{de_1}{dt}$ into the first equation of the system. Using Derive to do this, **Author** the command

$$\boxed{\text{DIF(\#13,t) = 0.1(\#13) - 0.01e + 2}} \ .$$

The variable e is used in place of the function E2(t). Simplify and soLve for e to get the following display with truncated expressions:

$$
14: \quad \frac{d}{dt}\left[e^{3\,t\,/\,50}\left[c1\ \cos\left[\frac{\sqrt{14}\,t}{100}\right] + c2\ \sin\left[\frac{\sqrt{14}\,t}{100}\right]\right] - 12\right] = 0.1\left[e^{3\,t\,/\,50}\right.
$$

$$
15: \quad e^{3\,t\,/\,50}\left[\left[\frac{3\ c1}{50} + \frac{\sqrt{14}\ c2}{100}\right] \cos\left[\frac{\sqrt{14}\,t}{100}\right] + \frac{\sqrt{2}\ (3\ \sqrt{2}\ c2 - \sqrt{7}\ c1)\ \sin\left[-\right.}{100}\right.
$$

$$
16: \quad e = e^{3\,t\,/\,50}\left[\sqrt{2}\ (2\ \sqrt{2}\ c1 - \sqrt{7}\ c2)\ \cos\left[\frac{\sqrt{14}\,t}{100}\right] + \sqrt{2}\ (\sqrt{7}\ c1 + 2\ \sqrt{2}\ c2)\ SI\right.
$$

The solution for e_2 after issuing Expand and approX is

$$
18: \quad e = e^{0.06\,t}\ ((4\ c1 - 3.74165\ c2)\ \cos\ (0.0374165\ t) +
$$

$$
(3.74165\ c1 + 4\ c2)\ \sin\ (0.0374165\ t)) + \blacksquare\blacksquare
$$

To find the arbitrary constants c_1 and c_2 that satisfy the initial conditions, just use the expressions for E1(t) and E2(t) (#13 and #18) and place the two equations in the vector form with $e_1(0) = 250$ and $e_2(0) = 114$ by using the Manage Substitute command. soLve and approX. The display of the steps to do this is as follows:

$$19: \left[e^{3t/50} \left[c_1 \cos\left[\frac{\sqrt{14}\,t}{100}\right] + c_2 \sin\left[\frac{\sqrt{14}\,t}{100}\right] \right] - 12 = 250,\ e^{0.06\,t} \quad ((4\,c_1$$

$$20: \left[e^{3\cdot0/50} \left[c_1 \cos\left[\frac{\sqrt{14}\cdot0}{100}\right] + c_2 \sin\left[\frac{\sqrt{14}\cdot0}{100}\right] \right] - 12 = 250,\ e^{0.06\cdot0} \quad ((4\,c_1$$

$$21: \left[c_1 = 262,\ c_2 = \frac{20280000}{74833} \right]$$

$$22: [c_1 = 262,\ c_2 = 271]$$

Once the solutions are verified, they can be plotted. Highlight the right-hand side of the function definitions for e_1 and e_2. Then the Plot command and Overlay location are selected to open the plotting window. Before the plotting is started, use Scale to set x-scale to 1.5 and y-scale to 500. Then move the movable cross with the right direction arrow ($\rightarrow$) or the Move command. The Center command is executed to move the plot range to positive t values. Now, the Plot command is issued for both components of the solution to obtain the following graph:

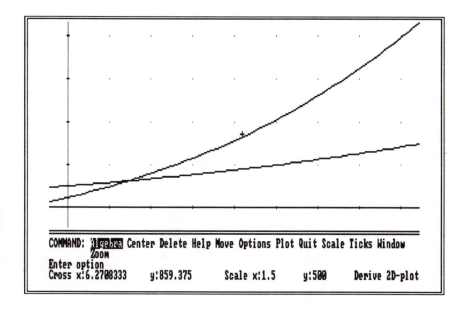

Therefore, even though procurement is currently the biggest expense, the organization would do better to think about long-range cost controls for its operating expenditures.

Exercise 5.10: Inventory and Pricing

The best mathematics is serious as well as beautiful—"important" if you like.

—G.H. Hardy [1940]

Subject: Analysis of Inventory and Pricing

Purpose: To solve a nonhomogeneous differential equation that models the pricing policy of a manufacturer.

References: Sections 1.10 and 1.13

Given: Based on several assumptions, a manufacturing company has developed simple differential models that relate price change, inventory level, production rate, and sales rate. The pricing policy for an item manufactured by the company is dependent on its inventory level. When the inventory is too high the price decreases, and when the inventory is too low the price rises. The simple linear model for this situation is

$$\frac{dp}{dt} = -\mu(L(t) - L_0),$$

where $p(t)$ is the forecast price, $L(t)$ is the inventory level at time t months, L_0 is the desired inventory level as constrained by their warehouse space, and μ is a small (usually ≤ 10), positive constant of proportionality that represents how tightly the inventory level is controlled.

Ultimately, the rate of change of the inventory depends on the production rate $Q(t)$ and the sales rate $S(t)$, which gives the equation

$$\frac{dL}{dt} = Q(t) - S(t).$$

$Q(t)$ and $S(t)$ can be modeled through their dependence on the price and its change by

$$Q(t) = a - bp - c\frac{dp}{dt}$$

and

$$S(t) = \alpha - \beta p - \delta\frac{dp}{dt},$$

where a, b, c, α, β, and δ are constants.

Exercises:

1. Use the given equations to write a second-order differential equation for p in terms of t and the constants.

2. The marketing department has determined the following values for the constants and initial conditions: $a = 0.2$, $b = 1/4$, $c = 0.7$, $\alpha = 14$, $\beta = 3$, $\delta = 2$, $p(0) = 100$, and $p'(0) = 0$. Set up and solve the differential equation in #1 in terms of the parameter μ, using the appropriate commands from the utility file ODE2.MTH.

3. Plot $p(t)$ for $\mu = 0.1$ and $\mu = 1$ on the same axes, showing accurate plots for $0 < t < 3$.

4. Based on the results shown in these plots, determine whether tight inventory control (larger value of μ) or loose control (smaller μ) will result in a lower price for the product at $t = 3$.

5. Use the equation for the sales rate $S(t)$ and the solutions for the two values of μ in #4 to determine which of these two values of μ results in the most total sales over the 3-month period. (**Hint:** Let Derive perform the integration.)

6. Ask yourself other questions related to this problem that you would like to investigate. Determine if Derive is a proper tool to use in this investigation by trying to answer your questions.

Example 5.11: Laplace Transform

Curiouser and curiouser!

—Lewis Carroll, *Alice in Wonderland* [1865]

Subject: Using Laplace Transforms to Solve Differential Equations with Forcing Functions Defined Piecewise

References: Sections 1.6, 1.10, and 1.12

Problem: Find the solution to the initial-value problem

$$O(y) := y'' + 4y = u(t) = \begin{cases} 4t & \text{if } 0 \le t < 1 \\ 4 & \text{if } t \ge 1 \end{cases},$$

with $y(0) = 1$ and $y'(0) = 0$.

Solution: First use the STEP function to define the piecewise forcing function $u(t)$. This can be done with the Declare Function command with function name input as u and definition as $\boxed{\text{STEP(a-t)b + STEP(t-a)c}}$. This produces the Derive function

$$u(a, b, c, t) = \begin{cases} b & \text{if } t < a \\ c & \text{if } t > a \end{cases}.$$

This piecewise function could be continuous (as it is in this problem) or discontinuous, depending on the values of b and c. The forcing function for this problem is entered as $\boxed{\text{u(1,4t,4,t)}}$, and the following display results:

```
1:   U (a, b, c, t) := STEP (a - t) b + STEP (t - a) c

2:   U (1, 4 t, 4, t)
```

It should be noted that there are often many alternative ways to enter functions like u into Derive. The method of using Derive's programming constructs, like the IF command, does not work well for functions that will be manipulated (such as taking derivatives) like u will be used. A more direct and alternate method to define the function u still using the STEP function would have been to enter

$$\boxed{\text{u(t):= 4t STEP(1-t) + 4 STEP(t-1)}}.$$

The plot of u using the Plot Overlay Zoom Both Out Plot sequence of commands reveals its stepped behavior. The resulting Derive plot is shown.

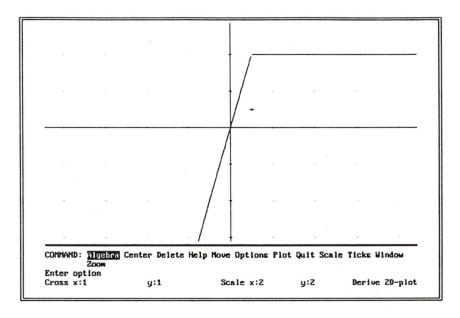

To find the Laplace transform $L(u(t))$, the formula

$$L(u(t)) = \int_0^\infty e^{-st} u(t)\, dt, \quad s > 0,$$

can be used. To implement this with Derive, first declare s positive with the Declare Variable command. Then the integrand is formed using the Author command and the following input:

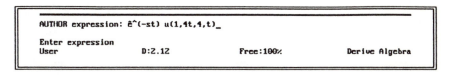

The menu command Calculus Integrate is used to obtain the transform. The user is prompted for the variable of integration: t; the lower limit: 0; and the upper limit: inf. Simplify and the result is as shown.

$$3: \quad \hat{e}^{-s\,t} \; U\,(1,\,4\,t,\,4,\,t)$$

$$4: \quad \int_{0}^{\bullet} \hat{e}^{-s\,t} \; U\,(1,\,4\,t,\,4,\,t)\,dt$$

$$5: \quad \frac{4}{s^2} - \frac{4\,\hat{e}^{-s}}{s^2}$$

This integral function is already defined in the utility file INT_APPS.MTH as the command LAPLACE(u,t,s). Using this command is convenient, if several transforms or other functions in the INT_APPS file are needed. In that case, we can load the utility file and use its functions by selecting the **Transfer Load Utility** command.

As an illustrative example, we will load this file to check our previous transform. Select the **Transfer Load Utility** command and enter INT_APPS for the filename. Author the command LAPLACE(#2,t,s) . Execute Simplify. The result confirms the results of our integration to find the Laplace transform of our forcing function.

By taking the Laplace transform of the left side of the equation by hand, the transformed equation becomes

$$s^2 Y(s) - y'(0)s - y(0) + 4Y(s) = \frac{4 - 4e^{-s}}{s^2}.$$

Author this equation with the initial conditions substituted for $y'(0)$ and $y(0)$ by entering

```
AUTHOR expression: s^2Y-1+4Y=4 / s^2 - 4 ê^(-s) / s^2_

Enter expression
Simp(4)              D:2.12            Free:100%           Derive Algebra
```

Then, so**L**ve for $Y(s)$ to obtain the following screen image:

$$9: \quad y = \frac{1}{s^2} - \frac{4\,\hat{e}^{-s}}{s^2\,(s^2 + 4)}$$

In order to find the inverse Laplace transform, use the **Expand** command to get the partial fraction decomposition form of the forcing function on the right-hand side of the equation as shown. Just press **Enter** or s **Enter** when queried for the EXPAND variable.

$$10: \quad y = \frac{\hat{e}^{-s}}{s^2 + 4} - \frac{\hat{e}^{-s}}{s^2} + \frac{1}{s^2}$$

Unfortunately, Derive does not provide a direct way to find the inverse transform. Derive does not do the contour integration necessary to find inverse Laplace transforms by means of the definition. Neither does Derive keep a table of values for this operation, so it must be done by hand from tables. This is not difficult for this problem once the partial fraction decomposition is performed. The solution is

$$y(t) = \frac{1}{2}\sin(2t-2)\text{STEP}(t-1) - (t-1)\text{STEP}(t-1) + \cos(2t) - \frac{1}{2}\sin(2t) + t.$$

Before plotting the solution, let's check to make sure it solves the equation and initial conditions. First, define the differential operator $O(y)$ for this differential equation by

```
AUTHOR expression: o(y):=dif(y,t,2) +4y _

Enter expression
Expd(?)              D:2.12           Free:100%        Derive Algebra
```

Then, **Author** the solution by

```
AUTHOR expression: y:=.5sin(2t-2)step(t-1)-(t-1)step(t-1)+cos(2t) -.5sin(2t)+t

Enter expression
User                 D:2.12           Free:100%        Derive Algebra
```

Author the expression $\boxed{O(y)}$ and **Simplify** to obtain the value of the operator, as shown in the following figure:

```
9:   O (y) :=  [d/dt]² y + 4 y

10:  y := 0.5 SIN (2 t - 2) STEP (t - 1) - (t - 1) STEP (t - 1) + COS (2 t) - 0.

11:  O (y)

12:  2 (t + 1) - 2 |t - 1|
```

Compare this expression with the right side of the equation $u(t)$ by *Authoring* their difference and executing **Simplify**. Just as we hoped, the result indicates the solution is correct.

```
13:  2 (t + 1) - 2 |t - 1| - U (1, 4 t, 4, t)
14:  0
```

Check the initial condition $y(0) = 1$ by using the Manage Substitute command to replace t with 0 in the expression for the solution. This result further verifies the solution.

```
16:  0.5 SIN (2 t - 2) STEP (t - 1) - (t - 1) STEP (t - 1) + COS (2 t) - 0.5 SIN
17:  0.5 SIN (2 0 - 2) STEP (0 - 1) - (0 - 1) STEP (0 - 1) + COS (2 0) - 0.5 SIN
18:  1
```

Check the initial condition $y'(0) = 0$ by highlighting the solution and executing Calculus Differentiate with respect to t, and Simplify. Use the Manage Substitute command to substitute 0 for t, and Simplify. All these operations are shown in the following figure:

```
        d
19:  --- (0.5 SIN (2 t - 2) STEP (t - 1) - (t - 1) STEP (t - 1) + COS (2 t) - 0.5
        dt

      ⌈ COS (2 t - 2)    1 ⌉                 COS (2 t - 2)
20:   | -----------  -  --- | SIGN (t - 1) + -----------  - COS (2 t) - 2 SIN
      ⌊      2           2 ⌋                       2

      ⌈ COS (2 0 - 2)    1 ⌉                 COS (2 0 - 2)
21:   | -----------  -  --- | SIGN (0 - 1) + -----------  - COS (2 0) - 2 SIN
      ⌊      2           2 ⌋                       2

22:  0
```

Now, the solution can be plotted. First, highlight the solution, and then execute the Plot Overlay command. Make sure to execute Delete All in order to delete any previous plots. Then execute Plot. This graph gives a nice visualization of the local behavior of the solution near the origin.

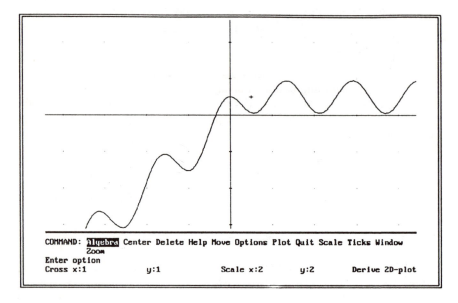

In order to obtain a plot showing more global behavior, execute the command Zoom Both Out twice. The resulting plot is as follows:

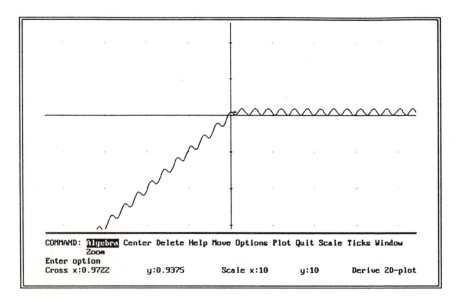

Exercise 5.12: Evaluating Laplace Transforms

The purpose of computing is insight, not numbers.

—R. W. Hamming

If you meet with any obstacles or difficulties, or are retarded with any doubts while you are walking on this cumbersome road of this study of mathematics, I beg you to impart them, and I shall endeavor to remove every hindrance out of your way to the best of my knowledge and ability.

—Isaac Barrow [1664]

Subject: Evaluating Laplace Transforms Using a Utility File

Purpose: To use Derive to find the Laplace transform of several types of functions using the integral definition in the utility file INT_APPS.MTH.

References: Sections 1.10 and 1.12

Given: The following functions, $f_i(t)$:

$$f_1(t) = \sin 3t \cos 3t,$$

$$f_2(t) = \begin{cases} -1, & \text{if } 0 \le t < 2 \\ 1, & \text{if } t \ge 2 \end{cases},$$

$$f_3(t) = \sin^2 t,$$

$$f_4(t) = e^{-t} \sin^2 t,$$

$$f_5(t) = \begin{cases} \cos t, & \text{if } 0 \le t < \pi \\ t, & \text{if } t \ge \pi \end{cases},$$

$$f_6(t) = e^{-t} t \cos 2t,$$

$$f_7(t) = \frac{\sin 3t}{t},$$

$$f_8(t) = 4e^{t^2}.$$

Exercises:

1. Plot each of the eight functions in the interval $0 < t < 10$. Think of them as forcing functions of a mechanical system. Which are periodic, discontinuous, smooth?

2. Load the Utility file INT_APPS.MTH into the work space. Use the function LAPLACE(u,t,s) to attempt to calculate the Laplace transform of these eight functions. Which functions can't it evaluate? Can you do the transforms more quickly by hand using a table?

3. From this test case of eight functions, is it worth your time to use Derive to do Laplace transforms?

4. Take the limit of each of the resulting transforms as $s \to \infty$. What do you believe is happening to these functions based on these results?

Exercise 5.13: Electrical Circuit Solved by Laplace Transforms

Begin at the beginning ... and go on till you come to the end: then stop.

—Lewis Carroll

Subject: Analysis of a Series Electrical Circuit Using Laplace Transforms

Purpose: To solve a second-order, nonhomogeneous model for a simple-loop electrical circuit using the techniques of Laplace transforms, with Derive performing many of the messy manipulations.

References: Sections 1.12, 5.11, and 5.12

Given: In Example 5.6, a differential model for a simple-loop electrical circuit with a time-dependent voltage source $(E(t))$, an inductor (L), a resistor (R), and a capacitor (C) was constructed. Letting $q(t)$ be the time-dependent charge on the capacitor, the model for such a circuit is

$$L\frac{d^2q}{dt^2} + R\frac{dq}{dt} + \frac{1}{C}q = E(t).$$

For this problem, the voltage source is defined by $E(t) = \alpha\cos 2t$, for $0 < t < \pi/4$, and 0, thereafter. For a circuit with $L = 10$ henrys, $R = 50$ ohms, $C = 1/60$ farads, no initial charge on the capacitor $(q(0) = 0)$, and no initial current $(q'(0) = 0)$, the equation to be solved is

$$10\frac{d^2q}{dt^2} + 50\frac{dq}{dt} + 60q = E(t),$$

with $q(0) = 0$ and $q'(0) = 0$.

Exercises:

1. Write an expression for voltage source $E(t)$ using Derive's unit step function STEP(t).

2. Plot $E(t)$ by accurately showing the interval of $0 < t < 2$ when $\alpha = 110$.

3. Use Derive to find the Laplace transform of $E(t)$. (Note: Either set up the integral definition of a Laplace transform or use the LAPLACE command in the utility file INT_APPS.MTH. Don't forget to establish the proper declaration for the transform variable s.)

4. Determine the algebraic equation resulting from taking the Laplace transform of the differential equation.

5. Solve and expand the algebraic equation in #4 for Y. Take the inverse Laplace transform of Y using a table to find the solution $q(t)$ of the differential equation.

6. Find the equation for the current $i(t)$ from $q(t)$ using Derive.

7. Plot $i(t)$ for $\alpha = 110$ and $\alpha = -110$. What is the difference between these graphs? Use the movable cross to approximate the maximum absolute value of the current.

8. Ask yourself other questions related to this problem that you would like to investigate. Determine if Derive is a proper tool to use in this investigation by trying to answer your questions.

Example 5.14: Bessel Equation

The most beautiful thing we can experience is the mysterious.

—Albert Einstein [1930]

Subject: Using Bessel Functions to Solve Bessel Equations

Reference: Section 1.10

Problem: Two frequently occurring differential equations in science and engineering are the Bessel equation of order n and the parametric Bessel equation of order n. The two equations are given respectively by

$$x^2 y'' + xy' + (x^2 - n^2)y = 0$$

and

$$x^2 y'' + xy' + (\lambda^2 x^2 - n^2)y = 0.$$

Some of the best-known applications of these equations are the determination of oscillations of a hanging chain or string, the theory of vibrations of a circular membrane (a drum), the study of planetary motion, and the study of wave propagation, diffusion, and potential involving cylindrical geometries.

Derive provides capabilities to evaluate, analyze, and plot solutions to these two equations.

Solution: The functions that solve the two Bessel equations are, appropriately enough, called Bessel functions. By using the power series solution technique, a series solution that converges for $n > 0$ and $0 \leq x < \infty$ can be found. This solution is denoted by $J_n(x)$ and is called the Bessel function of the first kind of order n. A second linearly independent series solution can be found and defined in a related fashion. It is denoted by $Y_n(x)$ and called the Bessel function of the second kind of order n. Therefore, the general solution of the Bessel equation of order n is $y = aJ_n(x) + bY_n(x)$, where a and b are arbitrary constants.

Derive contains functional definitions and approximations for these two functions in the utility file BESSEL.MTH. This file is loaded into the work area using the **Transfer Load Utility** command. The following table contains some of the commands in this file and explains their use:

Command	Function
BESSEL_J(n,z)	Uses the integral definition of $J_n(z)$ to determine values of the function.
BESSEL_J_SERIES (n,z,m)	Uses a truncated series to approximate $J_n(z)$; good for $z < 10$ with m increasing as z increases; excellent for $z < 1$ and m as small as 4.
BESSEL_J_ASYMP(n,z)	Uses the asymptotic expression to approximate $J_n(z)$; good for $z > 1$ and excellent for $z > 10$.
BESSEL_Y(n,z)	Uses BESSEL_J to find $Y_n(z)$.
BESSEL_Y_SERIES (n,z,m)	Uses a truncated series to approximate $Y_n(z)$; good for $z < 10$ with m increasing as z increases; excellent for $z < 1$ and m as small as 4.
BESSEL_Y_ASYMP(n,z)	Uses the asymptotic expression to approximate $Y_n(z)$; good for $z > 1$ and excellent for $z > 10$.
AI_SERIES(z,m)	Evaluates the Airy function (a close relative to $J_{1/3}(z)$), which solves Airy's equation: $y'' - xy = 0$
BI_SERIES(z,m)	Evaluates the BAiry function, which is the second linearly independent solution of the Airy equation.

Functions in utility file BESSEL.MTH.

First, let's compare values for the three different ways to obtain $J_n(x)$ using Derive commands for fixed values of n and x.

(1) If $n = 1$ and $x = 2$, then Author the command $\boxed{\text{BESSEL\_J(1,2)}}$ and Simplify to obtain

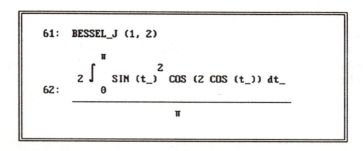

Unfortunately, Derive is unable to compute this definite integral exactly and just returns the set-up of the integral definition rather than its evaluation and issues a warning beep to alert the user that something is wrong. This means the symbolic integration for this function is too difficult. The question is, now what?

Luckily, there are several alternatives. The most obvious is probably to evaluate this integral definition with numerical quadrature. This is done through execution of the **Options Precision** command. Change the mode to **Approximate** by entering **A**, and then **Tab** over and enter the number of digits of accuracy desired. For this example, use 5. Now the integral can be approximated by re-execution of the command **Simplify**, except this time numerical quadrature will be performed. It takes several seconds to obtain the following result:

$$62: \quad \frac{2 \int_{0}^{\pi} SIN(t\_)^2 \; COS(2 \; COS(t\_)) \; dt\_}{\pi}$$

$$63: \quad 0.576639$$

(2) Another alternative method is to approximate $J_1(2)$ by the truncated series. This evaluation should be quite accurate for $x = 2$. Do this through **Author** and enter $\boxed{\text{BESSEL\_J\_SERIES}(1,2,m)}$. Try $m = 5$ first by putting 5 in for m directly in the command's input. Select **Simplify**. An answer close to that found by the integral approximation is obtained in much less time. The following Derive display shows the command and its result:

```
64:  BESSEL_J_SERIES (1, 2, 5)

65:  0.576724
```

Try $m = 10$ to see if the answer improves. It doesn't, so this value should be an accurate evaluation of $J_1(2)$.

(3) The final alternative is to use the asymptotic formula. Notice that the previous table indicates this technique may not be very accurate in this range where $x = 2$. Select Author and enter $\boxed{\text{BESSEL\_J\_ASYMP(1,2)}}$. Simplify and approX to get

```
66:   BESSEL_J_ASYMP (1, 2)

        √2 SIN (2)      √2 COS (2)
67:   ──────────── - ────────────
          2 √π            2 √π

68:   0.528775
```

This answer is a bit different (about 10% relative difference) from the previous approximations for $J_1(2)$. This seems to suggest that this asymptotic method may not be appropriate for small values of x (less than about 5).

Let's do the same experiment for the evaluation of $J_0(12.5)$. Author $\boxed{\text{BESSEL\_J(0,12.5)}}$ and select Simplify to obtain

```
69:   BESSEL_J (0, 12.5)

         π
         ⌠        ⎡ 25 COS (t_) ⎤
70:      ⎮  COS  ⎢ ─────────── ⎥ dt_
         ⌡        ⎣      2      ⎦
         0
      ──────────────────────────────
                     π

71:   0.146883
```

This calculation takes quite some time. The numerical quadrature has to work hard to get the 5-digit accuracy still required. Next try the truncated series approximations through the command $\boxed{\text{BESSEL\_J\_SERIES(0,12.5,m)}}$. Remember to replace m with the number of terms of the series to be used in the evaluation. Shown here are the commands and results for $m = 10$, 20, and 100. The calculation for $m = 100$ was performed in Exact mode and then approximated.

```
72:  BESSEL_J_SERIES (0, 12.5, 10)

73:  158.412

74:  BESSEL_J_SERIES (0, 12.5, 20)

75:  0.134974

76:  BESSEL_J_SERIES (0, 12.5, 100)

     30995508442065895746849572971813612166540570083001501558371209721247213014
80:  ────────────────────────────────────────────────────────────────────────────
     211020239775400444723575469283829006732164130830777503739235169595221145261

81:  0.146884
```

Only the calculation with the summation containing 100 terms is near the answer obtained with the previous integral method, which should be reasonably accurate. This indicates that the truncated series are inaccurate in this region of larger x unless many terms are kept in the series. The asymptotic approximation should do well for this value of x. The display for this operation is as follows:

$$82: \quad \text{BESSEL\_J\_ASYMP } (0,\ 12.5)$$

$$83: \quad \frac{2\,\text{SIN}\left[\dfrac{\pi}{4} + \dfrac{25}{2}\right]}{5\,\sqrt{\pi}}$$

$$84: \quad 0.148641$$

The asymptotic solution is close to the other two accurate solutions (integral and 100-term series) and is calculated in much less time. This experiment supports the table's suggestion that the asymptotic approximation is the most efficient technique for larger values of x.

Let's try plotting a Bessel function. The function we will use for this experiment is $J_{1/2}(x)$. Remember that the function evaluations for individual points for the three techniques (especially quadrature and truncated series) take quite some time. Therefore, a plotting window is opened so the plots are not redrawn, and even then the plotting will be slow.

First, open a plot window using the Window Split command. Select Horizontal and split at line #10. Move to the larger window by Window Next and execute Window Designate 2-D plot. Now the screen will keep the plot window open and the three functions can be plotted in sequence. Move back and forth from the algebra window to the plot window using the `Window Next` command or the `Algebra` and `Plot` commands.

Next, go back to the algebra window and `Author` the three functions (one at a time): i) `BESSEL_J(1/2,x)`, ii) `BESSEL_J_SERIES(1/2,x,3)`, and iii) `BESSEL_J_ASYMP(1/2,x)`. Highlight the first function and move back to the plot window. There is no need to plot values for $x < 0$; therefore, before plotting in the default window that centers the plot on the origin, use the direction keys or the `Move` command to move the cross to the right a couple of units, and then execute the `Center` command. This should change the horizontal plotting interval to show more of $x > 0$. Some finer tuning can be used to get just the positive x-axis to show in the window. When the interval is acceptable, use the `Plot` command to get the plot of the first highlighted function. (*Warning: Some of these plots take considerable time to produce.*) When the first plot is done, move back to the algebra window and highlight the next function. Then go back to the plot window and `Plot` that function. The display, after all three functions are plotted, looks like this:

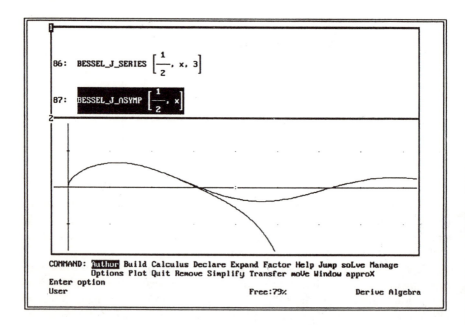

The 3-term series approximation doesn't do very well for larger x.

Similar evaluations, plots, and experiments can be done with the Bessel functions of the second kind $Y_n(x)$ and the Airy functions $Ai(x)$ and $Bi(x)$. However, we won't do them here.

Now, let's use the information we have learned about Derive's Bessel function generation to solve the following differential equation:

$$x^2 y'' + xy' + (x^2 - 0.01)y = 0,$$

with boundary conditions $y(1) = 3$ and $y(10) = 6$.

This is a Bessel equation of order 0.1, and the general solution is $y = aJ_{0.1}(x) + bY_{0.1}(x)$. The boundary conditions require the following two equations to be satisfied:

$$3 = aJ_{0.1}(1) + bY_{0.1}(1),$$
$$6 = aJ_{0.1}(10) + bY_{0.1}(10).$$

The two Bessel functions are evaluated using the four appropriate functions in the previous table, as shown in this Derive display.

```
88:   BESSEL_J_SERIES (0.1, 1, 25)

89:   0.770765

92:   BESSEL_Y_ASYMP (0.1, 1)

93:   0.0458707

94:   BESSEL_J_ASYMP (0.1, 10)

95:   -0.235487

96:   BESSEL_Y_ASYMP (0.1, 10)

97:   0.0905946
```

The two initial conditions are easily converted to a system of equations as components of a vector using the Author command to get

```
99:   [3 = a 0.770765 + b 0.0458707, 6 = a (-0.235487) + b 0.0905946]
```

The equations are solved for a and b using the commands soLve and approX. These commands produce the following result:

$$100: \left[a = -\frac{34404000000}{806291003999}, \; b = \frac{53310510000000}{806291003999} \right]$$

$$101: [a = -0.0426694, \; b = 66.1182]$$

So the solution to the equation is $y = -0.04266 J_{0.1}(x) + 66.1182 Y_{0.1}(x)$. This function can be *Authored* into the work area with the expression

```
-0.04266 BESSEL_J(0.1,x) + 66.1182 BESSEL_Y(0.1,x)
```
.

Then move into the plot window, clear all previous plots, and adjust the plot interval to show $5 < x < 9$ accurately and set the y-scale to 10. Then Plot the expression, whose graph appears as follows:

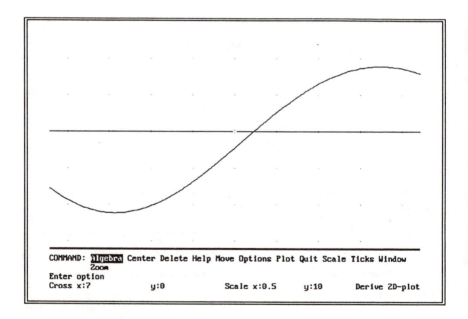

6

Matrix Algebra and Systems of Equations

To see the cooperation between calculus and linear algebra is to see the best parts of modern applied mathematics.

—Gilbert Strang [1986]

Sample problems have been worked out to show the power and versatility of Derive as a problem-solving tool in matrix algebra and systems of differential equations. In carefully reading these problems and working along with Derive, the reader should get a better feel for these subjects. Some of these examples also show the limitations of the computer. Some examples involve application problems, while others are posed in a more mathematical context. Many of the problems are similar to those typically found in undergraduate applied differential equations textbooks.

There are also exercises to be done. These problems direct you to learn new things about differential equations, Derive, and mathematics. Some of the problems' solution techniques and Derive commands are similar to those in the examples and refer to examples as a source of help. Other exercises lead you to explore the mathematics and software and to discover new results on your own. As you solve the questions from these exercises, ask yourself "what if" questions and attempt to answer them. Good luck, and have fun learning and exploring.

Example 6.1: Matrix Algebra

Mathematics is the science which draws necessary conclusions.

—Benjamin Peirce, *Linear Associative Algebra* [1870]

Subject: Solving Systems of Linear Algebraic Equations

References: Sections 1.6 and 1.8

Introduction: In problem solving, there are times when a solution to a linear system of algebraic equations is needed. There are several ways to solve systems of linear equations using Derive. The following examples demonstrate some of the ways to solve systems of equations along with a few extra features in Derive's matrix algebra capabilities.

Problem A: For the first example, let's find the solution to the 4×4 linear system

$$2x + y + 2z - 3w = 0,$$
$$4x + y + z + w = 15,$$
$$6x - y - z - w = 5,$$
$$4x - 2y + 3z - w = 2.$$

Solution to A: One way to solve this is to Author the four equations as components of a vector in Derive's work area. The expression is entered and displayed (truncated) as

```
AUTHOR expression: [2x+y+2z-3w=0,4x+y+z+w=15,6x-y-z-w=5,4x-2y+3z-w=2]_

Enter expression
                                    Free:100%              Derive Algebra
```

```
1:    [2 x + y + 2 z - 3 w = 0, 4 x + y + z + w = 15, 6 x - y - z - w = 5, 4 x -
```

Then, **soLve** this expression to obtain the solution

> 2:   [x = 2, y = 3, z = 1, w = 3]

The answer is produced, very conveniently, in less than one second.

Problem B: Let's solve the same system as in A using matrices with Derive.

Solution to B: If the above system is written as a matrix-vector system $A\vec{x} = \vec{f}$, then the solution vector $\vec{x}$ can be determined by $A^{-1}\vec{f}$. For this problem the solution is found by the evaluation of

$$\begin{bmatrix} 2 & 1 & 2 & -3 \\ 4 & 1 & 1 & 1 \\ 6 & -1 & -1 & -1 \\ 4 & -2 & 3 & -1 \end{bmatrix}^{-1} \begin{bmatrix} 0 \\ 15 \\ 5 \\ 2 \end{bmatrix}.$$

This expression is entered into the working area through the **Author** command by typing

> [[2,1,2,-3],[4,1,1,1],[6,-1,-1,-1], [4,-2,3,-1]]^(-1).[0,15,5,2]

Don't forget the dot (.) between the matrix and the vector. Derive uses the dot as the matrix multiplication symbol. Derive displays this expression as

> 3: $\begin{bmatrix} 2 & 1 & 2 & -3 \\ 4 & 1 & 1 & 1 \\ 6 & -1 & -1 & -1 \\ 4 & -2 & 3 & -1 \end{bmatrix}^{-1}$ · [0, 15, 5, 2]

and the **Simplify** command gives the answer

> 4:   [2, 3, 1, 3]

Problem C: These nice integer answers are typical of textbook problems, but such simple results are not expected very often in realistic problems. Not only that, but solutions to linear systems with integer coefficients can result in quite messy real numbers. For instance, let's change the last equation in the above system to $2x - 2y + 3z - w = 2/7$ and obtain the solution to this new system of equations.

Solution to C: To do this easily, highlight the vector of equations previously entered and use the Author command and **F3** key to bring that expression into the working area. The **Ctrl-S** and **Ctrl-D** keys provide movement of the cursor in the expression so the coefficients in the last expression can be changed. This is an instance when smart keystroking saves a lot of time. No one wants to retype all 4 of these equations. **Enter** this expression and soLve to obtain the solution in only a second or so, depending on the speed of the computer being used.

$$6: \quad \left[2\,x + y + 2\,z - 3\,w = 0,\ 4\,x + y + z + w = 15,\ 6\,x - y - z - w = 5,\ 2\,x - \right.$$

$$7: \quad \left[x = 2,\ y = \frac{361}{147},\ z = \frac{211}{147},\ w = \frac{457}{147} \right]$$

These rational numbers can be converted to approximate 6-digit decimal equivalents with the approX command to get

$$7: \quad \left[x = 2,\ y = \frac{361}{147},\ z = \frac{211}{147},\ w = \frac{457}{147} \right]$$

$$8: \quad [x = 2,\ y = 2.45578,\ z = 1.43537,\ w = 3.10884]$$

Problem D: Derive has no problem handling parameters in the equations. For instance, let's change the 2/7 in the last equation to the parameter α (it's time to see Derive do Greek letters).

Solution to D: Follow the keyboarding steps above to obtain the previous set of equations and change the 2/7 in the last equation to α. To get Derive to use the Greek letter α, type **Alt-a**. soLve this expression with the solve variables set to x, y, z, and w to obtain

```
10:  [2 x + y + 2 z - 3 w = 0, 4 x + y + z + w = 15, 6 x - y - z - w = 5, 2 x -

                 5 κ - 53           4 κ + 29         κ + 65
11:  [x = 2, y = - ――――――― ,   z = ――――――― ,   w = ――――― ]
                    21                 21             21
```

Problem E: This example problem shows when and how matrix manipulation can be used effectively to solve systems of equations. Sometimes problems require solutions for several systems with different right-hand sides. The following matrix equation can represent 4 different 3×3 systems of equations. In matrix form, the problem is to solve

$$
\begin{bmatrix} 1 & -1 & 1 \\ 1 & -2 & -2 \\ 2 & 1 & 3 \end{bmatrix}
\begin{bmatrix} x & u & m & r \\ y & v & n & s \\ z & w & p & t \end{bmatrix}
=
\begin{bmatrix} 3 & 1 & -6 & 3/2 \\ 0 & -1 & 1 & 0 \\ 4 & 3 & -9 & 11/2 \end{bmatrix}.
$$

Solution to E: Another way to enter matrices into Derive is through the `Declare Matrix` command, which prompts the user for matrix size (rows and columns) and the elements of the matrix. Let's enter the two matrices given the above containing numbers as separate working expressions in Derive using the `Declare Matrix` command. The first matrix has 3 rows and 3 columns. The second matrix has 3 rows and 4 columns. The result of the input of the elements in these matrices is the following screen display of the two matrices:

$$
12: \begin{bmatrix} 1 & -1 & 1 \\ 1 & -2 & -2 \\ 2 & 1 & 3 \end{bmatrix}
$$

$$
13: \begin{bmatrix} 3 & 1 & -6 & \dfrac{3}{2} \\ 0 & -1 & 1 & 0 \\ 4 & 3 & -9 & \dfrac{11}{2} \end{bmatrix}
$$

The solution matrix is obtained by multiplying the inverse of the 3×3 matrix times the 3×4 matrix. This can be accomplished by `Author` of the command `#n^(-1).#k`, where n and k are the numbers of the expression for these matrices in the work area. Don't forget the dot between the matrices.

For the above Derive screen, the command is $\boxed{\texttt{\#12\^{}(-1).\#13}}$. This displays
as

$$
14: \quad \begin{bmatrix} 1 & -1 & 1 \\ 1 & -2 & -2 \\ 2 & 1 & 3 \end{bmatrix}^{-1} \cdot \begin{bmatrix} 3 & 1 & -6 & \dfrac{3}{2} \\ 0 & -1 & 1 & 0 \\ 4 & 3 & -9 & \dfrac{11}{2} \end{bmatrix}
$$

`Simplify` this expression to obtain the solution matrix.

$$
15: \quad \begin{bmatrix} \dfrac{1}{2} & \dfrac{1}{2} & -1 & 2 \\ -\dfrac{9}{8} & \dfrac{1}{8} & 2 & \dfrac{3}{4} \\ \dfrac{11}{8} & \dfrac{5}{8} & -3 & \dfrac{1}{4} \end{bmatrix}
$$

Problem F: There is one other matrix operation to discuss, row reduction.
The Derive command to perform this operation is `ROW_REDUCE`. Let's use this
command to solve the matrix equation,

$$
\begin{bmatrix} 4 & 1 & 4 & -3 \\ 4 & -1 & 1 & -1 \\ 4 & -2 & -1 & -1 \\ 2 & -2 & 0 & -1 \end{bmatrix} \begin{bmatrix} w \\ x \\ y \\ z \end{bmatrix} = \begin{bmatrix} 1 \\ 11 \\ -5 \\ -12 \end{bmatrix}.
$$

Solution to F: To do this, enter the augmented matrix,

$$
\begin{bmatrix} 4 & 1 & 4 & -3 & 1 \\ 4 & -1 & 1 & -1 & 11 \\ 4 & -2 & -1 & -1 & -5 \\ 2 & -2 & 0 & -1 & 12 \end{bmatrix}
$$

into the work area using one of the two methods mentioned previously. Then
issue the command `ROW_REDUCE(#n)`, where n is the expression number that
contains the augmented matrix. The display showing the result is as follows:

16:
$$\begin{bmatrix} 4 & 1 & 4 & -3 & 1 \\ 4 & -1 & 1 & -1 & 11 \\ 4 & -2 & -1 & -1 & -5 \\ 2 & -2 & 0 & -1 & 12 \end{bmatrix}$$

17: ROW_REDUCE
$$\begin{bmatrix} 4 & 1 & 4 & -3 & 1 \\ 4 & -1 & 1 & -1 & 11 \\ 4 & -2 & -1 & -1 & -5 \\ 2 & -2 & 0 & -1 & 12 \end{bmatrix}$$

Execute Simplify to perform the row reduction. The resulting matrix is shown below.

18:
$$\begin{bmatrix} 1 & 0 & 0 & 0 & -\dfrac{23}{22} \\ 0 & 1 & 0 & 0 & -\dfrac{152}{11} \\ 0 & 0 & 1 & 0 & \dfrac{164}{11} \\ 0 & 0 & 0 & 1 & \dfrac{149}{11} \end{bmatrix}$$

Now the solution vector can be directly determined as

$$\begin{bmatrix} \dfrac{-23}{22} \\[2mm] \dfrac{-152}{11} \\[2mm] \dfrac{164}{11} \\[2mm] \dfrac{149}{11} \end{bmatrix}.$$

Exercise 6.2: Solving Systems of Linear Equations

> *You state this result and then try to mortify me by saying that it is easy to prove. Since I can't succeed in doing it I appeal to your good nature to help me out of this difficulty.*

—Hermite [1892]

Subject: Solving Systems of Linear Equations

Purpose: To use the operations of Derive to solve a system of linear equations.

References: Sections 1.8 and 6.1

Given: The following system of four equations in four unknowns and a parameter p:

$$2w + 5x + 3y - 7z = 4,$$

$$-3w + 2x - y + z = 7,$$

$$0.5w + 3x - 9y + 4z = 4,$$

$$2w - 3x - y + 2z = p.$$

Exercises:

1. Write the system in the matrix/vector form $\mathbf{R}\vec{x} = \vec{b}$.

2. Find the inverse of matrix $\mathbf{R}$.

3. Find the solution vector $\vec{x}$ by performing the operation $\mathbf{R}^{-1}\vec{b}$.

4. Find $\vec{x}$ using four equations as components of a vector, and execute soLve.

5. Find $\vec{x}$ using the ROW_REDUCE command.

6. Plot the four components of $\vec{x}$ as functions of p on the same axes.

7. Find the value of p that makes $z = 0$ in the solution.

Exercise 6.3: Solving Systems of Nonlinear Equations

It is not knowledge, but the act of learning, not possession but the act of getting there, which grants the greatest enjoyment.

—Carl Friedrich Gauss [1808]

Subject: Solving Systems of Nonlinear Equations Using Numerical Techniques

Purpose: To use the operations of Derive to plot functions and solve nonlinear equations and systems of nonlinear equations.

Reference: Section 1.11

Given: The following nonlinear equation and system of two nonlinear equations:

$$f(x) = e^x + x^4 - x^3 + 2x^2 + x - 2 = 0,$$
$$g_1(x, y) = 3xy - x + 6 = 0,$$
$$g_2(x, y) = 8y^2 - 2x^2 - 2x = 0.$$

Exercises:

1. Plot $f(x)$ in the interval $-6 < x < 6$. How many real roots are there in this interval?

2. Find the real roots of $f_1(x)$ in the above interval to 6 decimal places.

3. Does $f(x)$ have any real roots outside of the above interval? Why?

4. Find the two pairs of solutions of the system of equations ($g_1 = 0$ and $g_2 = 0$).

Example 6.4: Characteristic Values

Nothing makes a little knowledge as dangerous as examination time.

—Anonymous

Subject: Finding Characteristic Values and Characteristic Vectors (Eigenvalues and Eigenvectors)

References: Sections 1.8 and 1.11

Introduction: One of the common tasks in working with systems of linear equations is to find the characteristic values and characteristic vectors of a matrix. Derive can help tremendously in these tasks. In fact, there are several easy ways to use Derive, which are demonstrated in the following examples.

Problem A: The first task is to find the characteristic values and vectors of the rather messy 3×3 matrix **A**, where

$$\mathbf{A} = \begin{bmatrix} 3 & 2/9 & -1 \\ 2 & 2 & 3 \\ 1 & 1/6 & 9 \end{bmatrix}.$$

Solution to A: First, enter the matrix **A** into the work area either directly or by using the **Declare Matrix** command (see Section 1.8). Then, **Author** the expression **EIGENVALUES #n**, where n is the expression number for the matrix. The Derive display as a result of these steps for this problem is shown on the next page.

$$
1: \quad \begin{bmatrix} 3 & \dfrac{2}{9} & -1 \\ 2 & 2 & 3 \\ 1 & \dfrac{1}{6} & 9 \end{bmatrix}
$$

$$
2: \quad \text{EIGENVALUES} \quad \begin{bmatrix} 3 & \dfrac{2}{9} & -1 \\ 2 & 2 & 3 \\ 1 & \dfrac{1}{6} & 9 \end{bmatrix}
$$

Simplify to yield the solutions for the three eigenvalues using the default variable w.

$$
3: \quad \left[w = \frac{5}{3}, \ w = \frac{37}{6} - \frac{\sqrt{271}}{6}, \ w = \frac{\sqrt{271}}{6} + \frac{37}{6} \right]
$$

The commands to find the characteristic vectors (eigenvectors) for each of these eigenvalues are found in utility file VECTOR.MTH. Use the Transfer Load Utility command to load this file. Section 1.8 has descriptions for many of the commands in this file.

The characteristic vectors are found by solving the matrix/vector system $(\mathbf{A} - w\mathbf{I})\vec{x} = \vec{0}$ for $\vec{x}$, where w are the eigenvalues. To find the characteristic vector for a matrix $\mathbf{A}$ with an eigenvalue w known exactly with Derive, the command EXACT_EIGENVECTOR(A,w) can be used. When the eigenvalue was obtained through an approximation technique, the command APPROX_EIGENVECTOR(A,w) is used.

Since the eigenvalues for this matrix were determined exactly, Author the command

$$
\boxed{\text{EXACT\_EIGENVECTOR(\#n,w)}} \ ,
$$

where n is the statement number of the matrix; in this example, n is 1 and $w = 5/3$. This produces the display shown.

$$4: \quad \text{EXACT\_EIGENVECTOR} \quad \left[\begin{bmatrix} 3 & \dfrac{2}{9} & -1 \\ 2 & 2 & 3 \\ 1 & \dfrac{1}{6} & 9 \end{bmatrix}, \dfrac{5}{3} \right]$$

Select Simplify to determine the eigenvector. The resulting vector is as follows:

$$5: \quad [\ x1 = @1 \quad x2 = -6\ @1 \quad x3 = 0\ ]$$

Notice the value @1 in two of the vector components. Derive uses the values @n, n integer, to stand for arbitrary constants. So any value can be substituted for @1 to produce an eigenvector associated with the eigenvalue $5/3$.

The other two eigenvectors, for the eigenvalues $37/6 - \sqrt{271}/6$ and $37/6 + \sqrt{271}/6$, are computed using the same steps. The Derive expressions for $w = 37/6 - \sqrt{271}/6$ are as follows:

6:   EXACT_EIGENVECTOR $\left[\begin{bmatrix} 3 & \dfrac{2}{9} & -1 \\ 2 & 2 & 3 \\ 1 & \dfrac{1}{6} & 9 \end{bmatrix}, \dfrac{37}{6} - \dfrac{\sqrt{271}}{6}\right]$

7:   $\left[x1 = @2 \quad x2 = @2 \left[\dfrac{9\sqrt{271}}{85} - \dfrac{66}{85}\right] \quad x3 = @2 \left[\dfrac{97\sqrt{271}}{510} - \dfrac{1703}{510}\right]\right]$

Similarly, the displays for $w = 37/6 + \sqrt{271}/6$ are as follows:

8:   EXACT_EIGENVECTOR $\left[\begin{bmatrix} 3 & \dfrac{2}{9} & -1 \\ 2 & 2 & 3 \\ 1 & \dfrac{1}{6} & 9 \end{bmatrix}, \dfrac{37}{6} + \dfrac{\sqrt{271}}{6}\right]$

9:   $\left[x1 = @3 \quad x2 = -@3 \left[\dfrac{9\sqrt{271}}{85} + \dfrac{66}{85}\right] \quad x3 = -@3 \left[\dfrac{97\sqrt{271}}{510} + \dfrac{1703}{510}\right]\right]$

Problem B: The next example problem is to find the characteristic values and vectors of a 4×4 matrix $\mathbf{B}$, where

$$\mathbf{B} = \begin{bmatrix} 1 & 2 & 3 & 1 \\ 2 & -1 & 4 & 2 \\ 3 & 4 & -1 & -2 \\ 4 & 2 & -2 & 6 \end{bmatrix}.$$

Solution to B: Derive may have difficulty working with 4×4 matrices because the characteristic polynomial is fourth degree. Some fourth-degree polynomials don't lend themselves to easy solutions. This particular matrix is nice in this regard. Derive doesn't mind complex numbers for eigenvalues. See Exercise 6.5 for a problem with complex eigenvalues. The moral of the story is that even systems as small as 4×4 can be very difficult.

After entering the matrix, the eigenvalues can be solved for by use of the CHARPOLY function and soLve menu command or the EIGENVALUES function. For some 4×4 matrices, Derive may work a long time (2–4 minutes) to get a solution or even give up completely. Derive indicates it has given up by returning the same expression it started with. The statement and results of using the EIGENVALUES and Simplify commands for this problem are shown.

$$
11: \quad \text{EIGENVALUES} \begin{bmatrix} 1 & 2 & 3 & 1 \\ 2 & -1 & 4 & 2 \\ 3 & 4 & -1 & -2 \\ 4 & 2 & -2 & 6 \end{bmatrix}
$$

$$
12: \quad \left[w = -\frac{3^{3/4} \sqrt{153 \sqrt{3} - 448 \cos\left[\dfrac{\text{ATAN}\left[\dfrac{93 \sqrt{7556040055974927}}{2288874709} \right]}{3} + \dfrac{\pi}{6} \right]}}{12} \right.
$$

In order to produce workable numbers for the resulting eigenvalues, the exact eigenvalues are approximated to 6-digit accuracy by executing approX. The results are

$$
13: \quad [w = -5.96604, \; w = 7.51137, \; w = -1.66509, \; w = 5.11976]
$$

Since the eigenvalues for this matrix were determined approximately, to find the eigenvector for **B** corresponding to eigenvalue -5.96604, Author the command

APPROX_EIGENVECTOR(#n,-5.96604) ,

where n is the statement number of the matrix; in this example, n is 10. This produces the display shown.

$$14:\ \text{APPROX\_EIGENVECTOR}\ \left[\begin{bmatrix} 1 & 2 & 3 & 1 \\ 2 & -1 & 4 & 2 \\ 3 & 4 & -1 & -2 \\ 4 & 2 & -2 & 6 \end{bmatrix},\ -5.96604\right]$$

Select approX to determine the eigenvector. The resulting vector is as follows:

$$15:\ [-0.169994,\ -0.619077,\ 0.713895,\ 0.279639]$$

Notice there are no values like @1 in the vector components. This is because the vector was determined by a numerical scheme.

The other three eigenvectors can be determined with the same procedure. The display of these computations for $w = 7.51137$ is as follows:

16: APPROX_EIGENVECTOR $\left[\begin{bmatrix} 1 & 2 & 3 & 1 \\ 2 & -1 & 4 & 2 \\ 3 & 4 & -1 & -2 \\ 4 & 2 & -2 & 6 \end{bmatrix}, 7.51137 \right]$

17: [0.213745, 0.260360, -0.0234714, 0.941262]

Problem C: The final problem in this section demonstrates one of the things that can happen when there are repeated eigenvalues. Let's work with the 3×3 matrix **C**, where

$$C = \begin{bmatrix} 1 & 1 & 1 \\ 2 & 1 & -1 \\ -3 & 2 & 4 \end{bmatrix},$$

and find its eigenvalues and eigenvectors.

Solution to C: Finding the eigenvalues produces the surprise that there is only one, not three as might be expected. Once again, by default Derive labels it w as shown in this computation:

$$19: \text{EIGENVALUES} \begin{bmatrix} 1 & 1 & 1 \\ 2 & 1 & -1 \\ -3 & 2 & 4 \end{bmatrix}$$

$$20: \ [w = 2]$$

To find the eigenvector, Author EXACT_EIGENVECTOR(#18,2) .

$$21: \text{EXACT\_EIGENVECTOR} \left[\left[\begin{bmatrix} 1 & 1 & 1 \\ 2 & 1 & -1 \\ -3 & 2 & 4 \end{bmatrix}\right], 2\right]$$

Then, select Simplify to obtain the eigenvector $\vec{k}$.

$$22: \ [\ x1 = 0 \quad x2 = @4 \quad x3 = -@4 \ ]$$

This is the only linearly independent eigenvector for this matrix. The next step is to compute the first generalized eigenvector by solving for $\vec{m}$ in $(\mathbf{C} - w\mathbf{I})\vec{m} = \vec{k}$. In Derive, let

$$\vec{m} = \begin{bmatrix} d \\ e \\ f \end{bmatrix}.$$

Author the expression $(\mathbf{C} - w\mathbf{I})\vec{m} = \vec{k}$ using the highlight and **F3** key where appropriate. **I** is the 3×3 identity matrix and is entered in Derive as IDENTITY_MATRIX(3). A specific value should be substituted for the arbitrary constant @4. Often the value 1 is the simplest to use, so let's use it with the Manage Substitute command. Then issue the Simplify and soLve commands to obtain the vector components.

$$
24: \quad \left[\begin{bmatrix} 1 & 1 & 1 \\ 2 & 1 & -1 \\ -3 & 2 & 4 \end{bmatrix} - 2 \; \text{IDENTITY\_MATRIX (3)} \right] \cdot [d, e, f] = [0, 1, -1]
$$

$$
25: \quad [-d + e + f = 0, \; 2\,d - e - f = 1, \; -3\,d + 2\,e + 2\,f = -1]
$$

$$
26: \quad [d = 1, \; e = @5, \; f = 1 - @5]
$$

Next, the second generalized eigenvector is computed as $\vec{p}$ in the matrix-vector equation $(\mathbf{C} - w\mathbf{I})\vec{p} = \vec{m}$. Let

$$
\vec{p} = \begin{bmatrix} g \\ h \\ i \end{bmatrix}.
$$

Just as the previous generalized eigenvector was found, Author the expression $(\mathbf{C} - w\mathbf{I})\vec{p} = \vec{m}$. Substitute 1 for the arbitrary constant @3. Then soLve to obtain the vector components.

$$
29: \quad \left[\begin{bmatrix} 1 & 1 & 1 \\ 2 & 1 & -1 \\ -3 & 2 & 4 \end{bmatrix} - 2 \; \text{IDENTITY\_MATRIX (3)} \right] \cdot [g, h, i] = [1, 1, 0]
$$

$$
30: \quad [-g + h + i = 1, \; 2\,g - h - i = 1, \; -3\,g + 2\,h + 2\,i = 0]
$$

$$
31: \quad [g = 2, \; h = @6, \; i = 3 - @6]
$$

When a system of differential equations has repeated eigenvalues that produce generalized eigenvectors, the solution takes on special form. See Examples 6.6 and 6.9 for more on the use of eigenvalues and eigenvectors in solving systems of differential equations.

Exercise 6.5: Finding Eigenvalues and Eigenvectors

I hear and I forget; I see and I remember; I do and I understand.

—Old Chinese Proverb

Subject: Finding Eigenvalues and Eigenvectors for Matrix Systems of Differential Equations

Purpose: To find the eigenvalues and eigenvectors of matrices and use them to determine the general solutions of homogeneous systems of differential equations.

References: Sections 1.8 and 6.4

Given: Three 3×3 systems of differential equations of the form $\vec{X}'(t) = \mathbf{A}_j \vec{X}(t)$, with $j = 1, 2, 3$, and

$$\mathbf{A}_1 = \begin{bmatrix} 0 & -1 & -1 \\ 3 & 3/4 & -3/2 \\ -1/2 & 1/8 & 1/4 \end{bmatrix},$$

$$\mathbf{A}_2 = \begin{bmatrix} 2 & 4 & 4 \\ -1 & -2 & 0 \\ -1 & 0 & -2 \end{bmatrix},$$

$$\mathbf{A}_3 = \begin{bmatrix} 2 & 2 & -1 \\ 1 & 0 & 0 \\ 0 & 1 & 0 \end{bmatrix}.$$

Exercises:

1. Find the eigenvalues for the three matrices, $\mathbf{A}_1$, $\mathbf{A}_2$, and $\mathbf{A}_3$.

2. From analysis of these eigenvalues, which system's solution will decay? Which will grow? Which will show periodic behavior? What is the period of the periodic behavior?

3. Solve for the independent eigenvectors of matrix $\mathbf{A}_2$ and write the general solution of the differential system in terms of complex numbers and complex variables (i.e., leave the imaginary number i in the eigenvalues and eigenvectors).

4. Simplify the solution in #3 in terms of functions of real numbers and real variables.

5. Plot the three components of the solution vector for #4 with

$$\vec{X}(0) = \mathbf{A}_2 = \begin{bmatrix} 1 \\ 1 \\ 1 \end{bmatrix}$$

on the same axes in the interval $-8 < t < 8$.

Example 6.6: System of Differential Equations

CAS is not just an aid to problem solving; students can be guided in the use of CAS to discover fundamental mathematical concepts for themselves.

—Nancy Baxter and Priscilla Laws [1989]

Subject: Solving a Nonhomogeneous System of Differential Equations Using Variation of Parameters

References: Sections 1.8 and 6.4

Problem: Find a general solution of

$$x_1' = x_1 - x_2 + \frac{e^{-t}}{1 + t^2},$$

$$x_2' = 2x_1 - 2x_2 + \frac{2e^{-t}}{1 + t^2},$$

with $x_1(1) = 0$ and $x_2(1) = 1$.

Solution: The matrix-vector form of the equation is

$$\vec{X}' = \begin{bmatrix} 1 & -1 \\ 2 & -2 \end{bmatrix} \vec{X} + \vec{F}(t),$$

where

$$\vec{X} = \begin{bmatrix} x_1 \\ x_2 \end{bmatrix} \quad \text{and} \quad \vec{F}(t) = \begin{bmatrix} \dfrac{e^{-t}}{1 + t^2} \\ \dfrac{2e^{-t}}{1 + t^2} \end{bmatrix}.$$

The first step is to solve the homogeneous system $\vec{X}' = \mathbf{A}\vec{X}$, where

$$\mathbf{A} = \begin{bmatrix} 1 & -1 \\ 2 & -2 \end{bmatrix}.$$

The eigenvalues of the matrix are found using Derive by `Author` of the in-line command

```
EIGENVALUES [[1,-1],[2,-2]]
```
,

which displays as

$$1: \quad \text{EIGENVALUES} \begin{bmatrix} 1 & -1 \\ 2 & -2 \end{bmatrix}$$

Select the command Simplify to obtain the Derive display

$$2: \quad [w = 0, w = -1]$$

which means that the two eigenvalues are $\lambda_1 = -1$ and $\lambda_2 = 0$.

The eigenvectors could be found using the EXACT_EIGENVECTOR command from utility file VECTOR.MTH. However, for this problem we determine the eigenvectors directly from their definition. Find the eigenvectors in the form $\begin{bmatrix} j \\ k \end{bmatrix}$ by solving

$$\begin{bmatrix} 1 - \lambda & -1 \\ 2 & -2 - \lambda \end{bmatrix} \begin{bmatrix} j \\ k \end{bmatrix} = 0.$$

For this problem, the solutions for j and k are simple. The calculations using Derive are done for $\lambda = -1$ and 0, respectively, by issuing Author and soLve (Derive uses w instead of λ). The resulting display is as follows:

$$3: \quad \begin{bmatrix} 1 & -1 \\ 2 & -2 \end{bmatrix} - w \text{ IDENTITY\_MATRIX } (2)$$

$$4: \quad \begin{bmatrix} 1 - w & -1 \\ 2 & -w - 2 \end{bmatrix}$$

$$5: \quad \begin{bmatrix} 1 - w & -1 \\ 2 & -w - 2 \end{bmatrix} \cdot [a, b]$$

6:     $[a (1 - w) - b, 2 a - b (w + 2)]$

7:     $[a (1 - 0) - b, 2 a - b (0 + 2)]$

8:     $[a = @2, b = @2]$

9:     $[a (1 - -1) - b, 2 a - b (-1 + 2)]$

10:    $[a = @3, b = 2 @3]$

Substituting 1 for j in both eigenvectors gives two linearly independent vectors, as shown:

$$\vec{K_1} = \begin{bmatrix} 1 \\ 2 \end{bmatrix} \text{ and } \vec{K_2} = \begin{bmatrix} 1 \\ 1 \end{bmatrix}.$$

From this result, the complementary solution is written as

$$\vec{X}_c = a \begin{bmatrix} 1 \\ 1 \end{bmatrix} + b \begin{bmatrix} 1 \\ 2 \end{bmatrix} e^{-t},$$

with a and b arbitrary constants, and the fundamental matrix, Φ, is

$$\begin{bmatrix} 1 & e^{-t} \\ 1 & 2e^{-t} \end{bmatrix}.$$

This matrix is used to solve for the particular solution by using the variation of parameters formula

$$\vec{X}_p = \Phi(t) \int \Phi(t)^{-1} \vec{F}(t)\, dt.$$

First, use Derive to find Φ^{-1}, by use of the **Author** command and by entering

$$\boxed{\texttt{[[1,Alt-e\^{}-t], [1,2Alt-e\^{}-t]]\^{}-1}}.$$

Notice that e is entered by pressing the **Alt** key and the letter **e** key, simultaneously. Derive displays it as ê. **Simplify** this expression to find Φ^{-1}. At this point, the Derive display and authored expression are as follows:

```
                   -t  -1
           ┌ 1   ê    ┐
   11:      │           │
           │        -t │
           └ 1  2 ê    ┘

           ┌  2    -1  ┐
           │  t    t   │
   12:     │  ê    ê   │
           └ - ê    ê  ┘
  ─────────────────────────────────────────────────────
  AUTHOR expression: [[2, -1], [- ê^t, ê^t]].[ê^-t/(1+t^2),2ê^-t/(1+t^2)]_

  Enter expression
  Solve(9)                          Free:89%              Derive Algebra
```

Multiply Φ^{-1} by $\vec{F}(t)$. This is best done by building from the previous expression for Φ^{-1} using the Author command and **F3**, then dotting this with $\vec{F}(t)$. Simplify this expression and integrate the resulting vector using the Calculus Integrate menu command. The integration must be done with respect to t with no explicit limit of integration specified. Just press Enter when queried for the limits. The resulting display upon the execution of the Simplify command is shown.

$$13: \begin{bmatrix} 2 & -1 \\ -\hat{e}^{t} & \hat{e}^{t} \end{bmatrix} \cdot \begin{bmatrix} \dfrac{\hat{e}^{-t}}{1+t^2}, & \dfrac{2\,\hat{e}^{-t}}{1+t^2} \end{bmatrix}$$

$$14: \begin{bmatrix} 0, & \dfrac{1}{t^2+1} \end{bmatrix}$$

$$15: \int \begin{bmatrix} 0, & \dfrac{1}{t^2+1} \end{bmatrix} \, dt$$

$$16: \quad [0, \text{ ATAN }(t)]$$

Following the equation for the variation of parameters given above, we multiply this expression by Φ to obtain $\vec{X}_\text{p}$, which is done by *Authoring* the Derive expression

```
AUTHOR expression: [[1, ê^(-t)], [1, 2 ê^(-t)]] · [0, ATAN (t)]_

Enter expression
User                                    Free:89%              Derive Algebra
```

Perform the operation through the command Simplify. The display for this is shown.

$$17: \quad \begin{bmatrix} 1 & \hat{e}^{-t} \\ 1 & 2\,\hat{e}^{-t} \end{bmatrix} \cdot [0, \text{ATAN (t)}]$$

$$18: \quad [\hat{e}^{-t}\ \text{ATAN (t), }2\,\hat{e}^{-t}\ \text{ATAN (t)}]$$

The general solution, $\vec{X}_g$, is $\vec{X}_c + \vec{X}_p$ and can now be written as

$$\vec{X}_g = a \begin{bmatrix} 1 \\ 1 \end{bmatrix} + b \begin{bmatrix} 1 \\ 2 \end{bmatrix} e^{-t} + \begin{bmatrix} 1 \\ 2 \end{bmatrix} e^{-t} \tan^{-1} t.$$

The arbitrary constants a and b are determined by substitution of the initial conditions into $\vec{X}_g$ and by solving for a and b. To do this with Derive, the two equations are entered as a vector of equations with x used for x_1 and y used for x_2. Then using the menu commands Manage Substitute, the initial values of $t = 1$, $x = 0$, and $y = 1$ are entered. The display showing these operations is given.

$$20: \quad [x = a + b\,\hat{e}^{-t} + \hat{e}^{-t}\ \text{ATAN (t), }y = a + b\,2\,\hat{e}^{-t} + 2\,\hat{e}^{-t}\ \text{ATAN (t)}]$$

$$21: \quad [0 = a + b\,\hat{e}^{-1} + \hat{e}^{-1}\ \text{ATAN (1), }1 = a + b\,2\,\hat{e}^{-1} + 2\,\hat{e}^{-1}\ \text{ATAN (1)}]$$

The values for a and b are obtained by selecting the command soLve. Derive displays the result as follows:

$$22: \quad \left[a = -1,\ b = \hat{e} - \frac{\pi}{4}\right]$$

Finally, the solution to the initial-value problem is written as

$$x_1 = -1 + (e - \pi/4)e^{-t} + e^{-t} \tan^{-1} t$$

and

$$x_2 = -1 + 2(e - \pi/4)e^{-t} + 2e^{-t} \tan^{-1} t.$$

To get a geometric perspective of this solution, x_1 and x_2 are plotted using Derive's 2-D plotting capability. Just Author the two solutions as equations into the Derive display, as shown.

$$24: \quad x = -1 + \left[\hat{e} - \frac{\pi}{4} \right] \hat{e}^{-t} + \hat{e}^{-t} \text{ ATAN } (t)$$

$$26: \quad y = -1 + \left[\hat{e} - \frac{\pi}{4} \right] 2 \hat{e}^{-t} + 2 \hat{e}^{-t} \text{ ATAN } (t)$$

Before plotting the two functions, remember to set the plotting parameters to those of your hardware using the Options Display menu command. A plot showing the local behavior near the origin is obtained with the scale of the tick marks set to the default value of 1.0 in both the x- (horizontal) and

y- (vertical) directions. This plot is as follows:

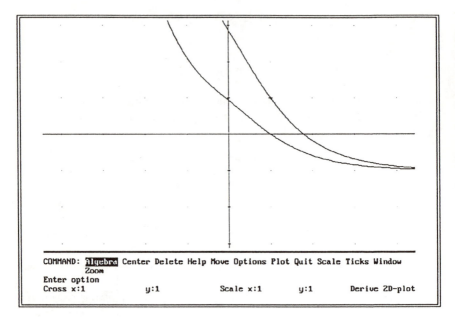

```
COMMAND: Algebra Center Delete Help Move Options Plot Quit Scale Ticks Window
         Zoom
Enter option
Cross x:1              y:1            Scale x:1        y:1        Derive 2D-plot
```

A plot of the global behavior is obtained by changing the scale to 10 for every tick mark on both axes. Change the scale by using the Scale command and entering 10 for both the x- and y-scale. Use the **Tab** key to move between these selections in the menu. This plot of the global behavior of the solution components is shown below.

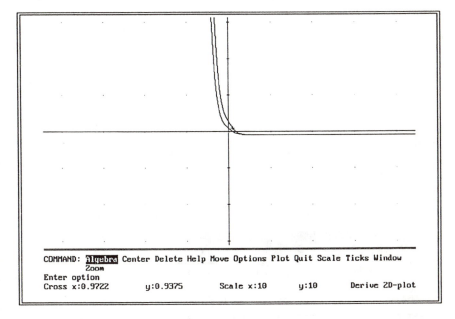

```
COMMAND: Algebra Center Delete Help Move Options Plot Quit Scale Ticks Window
         Zoom
Enter option
Cross x:0.9722      y:0.9375      Scale x:10      y:10        Derive 2D-plot
```

The window could be split horizontally to show both the algebraic and graphical forms of the solutions. This can be done automatically by selecting the Under location or by explicitly opening a plotting window. To do this, go back to the algebra screen by selecting Algebra. Execute the Window Split command and select a Horizontal split at line #13. Issue the Plot command. In order to make a better plot for this window, change the x-scale to 1 and the

y-scale to 25. The following screen image is the final result of these operations:

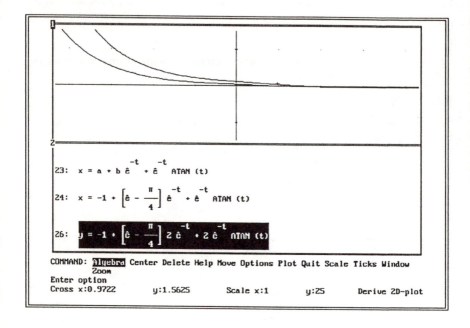

Exercise 6.7: Linear Differential System

Having to conduct my grandson through his course of mathematics, I have resumed the study with great avidity. It was ever my favorite one. We have no theories there, no uncertainties remain on the mind; all is demonstration and satisfaction.

—Thomas Jefferson [1811]

Subject: 3×3 Homogeneous, Linear Differential System with Constant Coefficients

Purpose: To find the solution of a 3×3 system of linear, homogeneous differential equations with an initial value through the determination of eigenvalues and eigenvectors.

References: Sections 1.8, 6.4, and 6.6

Given: The system of three differential equations

$$x_1' = 3x_1 - 2x_3,$$
$$x_2' = -x_1 + x_2 + x_3,$$
$$x_3' = 4x_1 - 3x_3,$$

with initial conditions

$$\vec{x}(0) = \begin{bmatrix} x_1(0) \\ x_2(0) \\ x_3(0) \end{bmatrix} = \begin{bmatrix} 3 \\ 1 \\ 3 \end{bmatrix}.$$

Exercises:

1. Write the given system of equations in vector/matrix form.

2. Find the eigenvalues of the characteristic matrix for this system.

3. Find an eigenvector for each of the three eigenvalues determined in #2.

4. Write the general solution, $\vec{x}_g(t)$, using a, b, and c as the arbitrary constants.

5. Use the given initial condition to determine the particular values for a, b, and c.

6. Graph the three component functions of $\vec{x}(t)$—x_1, x_2, and x_3—on one set of axes from $t = 0$ to $t = 10$. (**Hint:** You can plot these functions as components of a vector or on a split screen to save time.)

7. Which component function has the greatest value at $t = 10$?

Exercise 6.8: Interacting Species

Education is the instruction of the intellect in the laws of nature.

—Thomas Huxley [1868]

Subject: Linear Models for the Population of Interacting Species

Purpose: To build simple linear system models for the population change of interacting species and determine the solution of the system of differential equations.

References: Sections 1.8, 6.4, and 6.6

Given: Populations of indigenous wildlife have developed in an isolated desert park. There are only two known competing species, the jack mole and the prairie rat. Several assumptions have been made in order to build a differential-equation model for their populations. $J(t)$ is used for the expression over time of the jack mole population, and $P(t)$ is the expression for the prairie rat. Data collection has indicated both species have identical growth rates that depend only on the size of the population of their own species and identical competition rates that depend only on the size of the other species. If

$$\vec{X}(t) = \left[\begin{array}{c} J(t) \\ P(t) \end{array} \right],$$

a simple normalized model for this interaction, with the time t measured in years, can be written as

$$\vec{X}'(t) = \left[\begin{array}{cc} 1 & -1 \\ -1 & 1 \end{array} \right] \vec{X}(t).$$

Exercises:

1. Use Derive to solve for the eigenvalues and eigenvectors of this model. Write the general solution for this system.

2. Without knowing the initial conditions for the populations, determine the possible scenarios for future population changes for these two species. Under what condition could both populations survive? What is most likely to happen?

3. If the population sizes are known to be 16,000 jack moles and 15,500 prairie rats, how long will both species coexist?

4. Plot the two population curves from $t = 0$ until one population no longer exists. Use the movable cross in Derive's plotting window to approximate the time when the smaller population reaches half its original population.

5. A population of a predator species of coyotes that preys on both of the previous species is introduced into the area. The coyote population's growth rate in isolation is identical to that of the other two species. The coyotes prey on the other two species at half that growth rate. Use $C(t)$ to denote the coyote population at time t, and write a 3×3 matrix model for the population interaction of these three species.

6. Solve this 3×3 model for the general solution, if the coyote population size is 1000 when the other two population sizes are as given in #3.

7. Plot the populations of jack moles and prairie rats for the new model on the same axes as the populations from the previous model. What does the model predict the effect of introducing coyotes into this desert park will be on the other two populations?

8. Ask yourself other questions related to this problem that you would like to investigate. Determine if Derive is a proper tool to use in this investigation by trying to answer your questions.

Example 6.9: Differential System for Compartment Models

Teaching children to count is not as important as teaching them what counts.

—Anonymous

Subject: Solving a Differential System for Compartment Models in Biology Using Matrix Algebra

References: Sections 1.8, 6.4, and 6.6

Problem: Biological organisms often are composed of component parts called compartments. Examples are the organs in the body—heart, lungs, stomach, and liver—or the different parts of the blood—red blood cells, plasma, and white blood cells. Individual compartments may interact with one another, and it's that interaction that is modeled by a system of differential equations. We want to find the concentration of a given substance in each compartment.

By assuming constant-volume and well-mixed compartments, mass transport equations for each compartment can be derived. The differential model for a system with three compartments whose only exit to the outside environment is through the first compartment is written as

$$y_1' = -(k_{21} + k_{13} + k_{01})y_1 + k_{12}y_2 + k_{13}y_3 + b_1(t),$$
$$y_2' = k_{21}y_1 - (k_{12} + k_{32})y_2 + k_{23}y_3 + b_2(t),$$
$$y_3' = k_{31}y_1 + k_{32}y_2 - (k_{13} + k_{23})y_3 + b_3(t),$$

where the k_{ij} are constant transfer coefficients that describe the ease of flow of the substance between compartments i and j. For example, k_{23} is the transfer coefficient for flow from compartment 2 to compartment 3. A 0 in the subscript signifies the interface with the external environment. A transfer coefficient of 0 means no transfer of the substance, and a coefficient of 1 means completely unimpeded flow. The b_i are direct input/output functions caused by the manufacture or use of the substance within the compartment.

If the values for the transfer coefficients and input functions are given as follows:

$$
\begin{array}{lll}
k_{01} = 0.2 & k_{12} = 0.1 & k_{13} = 0.15 \\
k_{21} = 0.3 & k_{32} = 0.15 & k_{23} = 0.25 \\
k_{31} = 0.25 & & \\
\end{array}
$$

$$
\begin{array}{lll}
b_1 = at^2 + bt + c & b_2 = 0 & b_3 = 0.01 \\
\end{array}
$$

then the equation in matrix/vector form becomes

$$
\vec{y}' = \begin{bmatrix} -0.65 & 0.1 & 0.15 \\ 0.3 & -0.25 & 0.25 \\ 0.25 & 0.15 & -0.4 \end{bmatrix} \vec{y} + \begin{bmatrix} at^2 + bt + c \\ 0.0 \\ 0.01 \end{bmatrix}.
$$

This is best solved using Derive with the steps presented in Example 6.6. In outline form, these steps are: i) enter the matrix and find its eigenvalues with the commands EIGENVALUES and Simplify; use approX, if necessary; ii) find the eigenvectors by using the appropriate commands (EXACT_EIGENVECTOR or APPROX_EIGENVECTOR) from the utility file VECTOR.MTH; iii) form and enter into Derive the fundamental matrix, Φ; iv) Author the integral formula for the particular solution, $\vec{y}_p$, into Derive and Simplify; and v) use the result to Author the general solution, $\vec{y}_g$, and Manage Substitute the initial values to get the solution.

The results of using Derive to do the first three steps for this problem are shown in the following three figures. Some statements are truncated.

5: APPROX_EIGENVECTOR $\left[\begin{bmatrix} -0.65 & 0.1 & 0.15 \\ 0.3 & -0.25 & 0.25 \\ 0.25 & 0.15 & -0.4 \end{bmatrix}, -0.519498\right]$

6: [-0.150450, 0.757158, -0.635669]

7: APPROX_EIGENVECTOR $\left[\begin{bmatrix} -0.65 & 0.1 & 0.15 \\ 0.3 & -0.25 & 0.25 \\ 0.25 & 0.15 & -0.4 \end{bmatrix}, -0.0157311\right]$

8: [0.247343, 0.837222, 0.487729]

9: APPROX_EIGENVECTOR $\left[\begin{bmatrix} -0.65 & 0.1 & 0.15 \\ 0.3 & -0.25 & 0.25 \\ 0.25 & 0.15 & -0.4 \end{bmatrix}, -0.76477\right]$

10: [0.842835, -0.263220, -0.469408]

11: $\begin{bmatrix} -0.1505\,e^{-0.5195\,t} & 0.2473\,e^{-0.0157\,t} & 0.8428\,e^{0.7648\,t} \\ 0.7572\,e^{-0.5195\,t} & 0.8372\,e^{-0.0157\,t} & -0.2632\,e^{-0.7648\,t} \\ -0.6357\,e^{-0.5195\,t} & 0.4877\,e^{-0.0157\,t} & -0.4694\,e^{-0.7648\,t} \end{bmatrix}$

It turns out that Derive cannot perform the symbolic integration of the fourth step for this problem. When this happens, it is possible to use a numerical approximation method. See Chapter 4 for more information and examples of numerical techniques. Just because Derive couldn't do this problem does not mean that it cannot solve nonhomogeneous 3×3 systems. There are many such systems that Derive can solve.

Exercise 6.10: Variation of Parameters

If a little knowledge is dangerous, where is the man who has so much as to be out of danger?

—Thomas Huxley [1877]

Subject: Solving a Nonhomogeneous System of Differential Equations by Variation of Parameters

Purpose: To solve and analyze a system of three linear differential equations using the method of variation of parameters.

References: Sections 1.8, 6.4, 6.6, and 6.9

Given: The following 3×3 differential system with an initial condition:

$$\vec{X}(t)' = \begin{bmatrix} -3 & 2 & 0 \\ 2 & -3 & 0 \\ 0 & 0 & -5 \end{bmatrix} \vec{X}(t) + \begin{bmatrix} 3.25 \\ 6t \\ 2e^t \end{bmatrix},$$

$$\vec{X}(0) = \begin{bmatrix} 0 \\ 1 \\ -1.5 \end{bmatrix},$$

where

$$\vec{X}(t) = \begin{bmatrix} x(t) \\ y(t) \\ z(t) \end{bmatrix}.$$

Exercises:

1. Find the eigenvalues of the coefficient matrix for this system.

2. Find an eigenvector for each of the three eigenvalues.

3. Write the three linearly independent solution functions for the homogeneous part of the system of equations.

4. Form the fundamental matrix from the set of three solutions found in #3.

5. Use Derive to find a particular solution to the nonhomogeneous system using the formulas in the method of variation of parameters.

6. Find the general solution to the system of equations and evaluate the three arbitrary constants from the given initial condition.

7. Plot on the same axes the three components of the solution in the interval $0 < t < 6$.

8. Which component—$x(t)$, $y(t)$, or $z(t)$—shows the most rapid growth in this region?

9. What is the limiting value for the three components as $t \to \infty$?

Exercise 6.11: Biological System

Education is what you have left over after you have forgotten everything you have learned.

—Anonymous

Subject: Systems of Equations for the Flow of a Substance Through Organs of a Biological System

Purpose: To model and analyze flow through a biological system using a system of linear differential equations.

References: Sections 1.8, 6.4, 6.6, 6.8, and 6.9

Given: A schematic diagram for the fluid flow through a simple biological system made up of Organs A and B is shown.

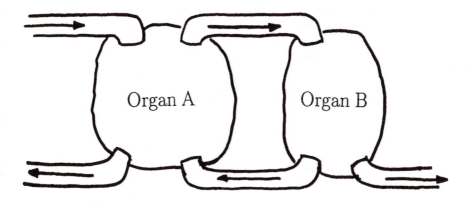

Organ A stays filled with 3 gallons of fluid, and Organ B maintains 2 gallons. The following fluid flows are established: external input to A is 0.05 gal/hour, external output from A is 0.03 gal/hour, flow from A to B is 0.05 gal/hour, flow from B to A is 0.03 gal/hour, and the external output of B is 0.02 gal/hour.

Exercises:

1. If the input into A is pure fluid (none of the tracking substance is present), write the differential equation model for the amount of the substance in Organ A (x_1) and Organ B (x_2).

2. If 3 oz of the substance is injected directly into Organ A and Organ B is free of the substance, solve the model for x_1 and x_2. Plot the equations for $0 < t < 80$.

3. When will the substance concentration in organ B exceed 0.1 oz/gal, and when will it fall below 0.1 oz/gal?

4. Model and solve for x_1 and x_2 if 3 oz of the substance is injected directly into Organ B instead of Organ A. Plot the components of the solution. When will the substance concentration in organ B fall below 0.1 oz/gal for these initial conditions?

5. Instead of the substance being injected all at once, it is introduced into the system through the input flow into A at a concentration of 6 oz/gal. Write the new model for this flow. How much of the substance is introduced into the system after 10 hours?

6. Solve the model in #5. Plot the solution. When will the substance concentration in Organ B exceed 0.1 oz/gal and fall below 0.1 oz/gal for this model?

7. Ask yourself other questions related to this problem that you would like to investigate. Determine if Derive is a proper tool to use in this investigation by trying to answer your questions.

Exercise 6.12: Series Electrical Circuit

... it behooves us to place the foundations of knowledge in mathematics.

—Roger Bacon [13th century]

Subject: Analysis of a Series Electrical Circuit Using Systems of Differential Equations

Purpose: To solve a nonhomogeneous system of two differential equations that model the current in a two-loop electrical circuit.

References: Sections 1.8, 5.6, 6.4, 6.6, and 6.8

Given: In Example 5.6, a differential model for a simple-loop electrical circuit with a voltage source $(E(t))$, an inductor (L), a resistor (R), and a capacitor (C) was constructed. Now a circuit with two loops in series needs to be modeled and analyzed. The schematic diagram for the circuit is given below.

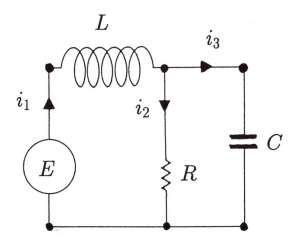

Letting $i_k(t)$, $k = 1$, 2, and 3, be the time-dependent current in the loops as shown, the model for this circuit is

$$L\frac{di_1}{dt} + Ri_2 = E(t),$$

$$RC\frac{di_2}{dt} + i_2 - i_1 = 0.$$

For this problem the voltage source is defined as a constant $E(t) = \alpha$ volts, $L = 0.5$ henrys, $R = 50$ ohms, $C = 0.00001$ farads, $i_1(0) = 0$ amps, and $i_2(0) = 0$ amps.

Exercises:

1. Write the model as a matrix/vector system.

2. Determine the eigenvalues, eigenvectors, and complementary solution for the homogeneous system.

3. Solve the nonhomogeneous system using variation of parameters subject to the given initial conditions.

4. If i_1 cannot exceed 2 amps, what is the maximum value for α?

5. Plot $i_1(t)$ and $i_2(t)$ on the same axes.

6. What is the long-term (steady-state) behavior of these two currents?

7. Ask yourself other questions related to this problem that you would like to investigate. Determine if Derive is a proper tool to use in this investigation by trying to answer your questions.

7

Partial Differential Equations

Just because a human cannot program a computer to think, does not mean that a computer cannot think.

—Anonymous

Sample problems involving partial differential equations have been worked out to show the power of Derive as a problem-solving tool for this topic. By reading these problems and working along with Derive, you should get a better feel for the subject. Some of these examples also show the limitations of the computer and may help you to know when and when not to use Derive. Some of the examples involve models from applications, while others are posed in a mathematical context. The problems are similar to those typically found in undergraduate differential-equations textbooks.

The exercises are to be done by you. These problems direct you to learn new things about differential equations. Some of the problems' solution techniques and Derive commands are similar to those in the examples, and others lead you to explore the mathematics and software and to discover new results on your own. As you solve the questions from these exercises, ask yourself questions and attempt to answer them.

Example 7.1: Fourier Series

> *Applied mathematics is alive and very rigorous. . . . Nothing is out of date about Fourier!*

> —Gilbert Strang [1986]

Subject: Fourier Series

References: Sections 1.7, 1.10, 1.11, and 1.12

Introduction: One of the common tasks often needed to solve partial differential equations is to expand functions in Fourier series. Derive can help in this task in several ways, as demonstrated in the following example problems.

Derive has the command FOURIER in the utility file for integral applications, INT_APPS.MTH, which produces Fourier series approximations to functions. The utility file must be loaded into the work space through execution of the **Transfer Load Utility** command before the FOURIER command can be used. The command FOURIER(y(t),t,t1,t2,n) produces the Fourier series approximation with n terms (both sine and cosine) to $y(t)$ from $t = t_1$ to $t = t_2$. The command evaluates the coefficients of the series through their integral definition.

Problem A: Just how many terms should be used to get a good approximation for a function? Let's try several different Fourier series approximations for a simple function like $y(x) = x^3$, $-3 < x < 3$.

Solution to A: For this problem, the independent variable is x (instead of the default of t used in the Introduction). To start with, let $n = 2$, 4, and 6. The sequence of input commands and outputs after executing **Simplify** are shown in the following Derive screen images:

```
AUTHOR expression: fourier(x^3,x,-3,3,2)

Enter expression
User                              Free:93%              Derive Algebra
```

12: FOURIER $(x^3, x, -3, 3, 2)$

13: $$\frac{54\,(\pi^2 - 6)\,\sin\left[\dfrac{\pi x}{3}\right]}{\pi^3} - \frac{27\,(2\pi^2 - 3)\,\sin\left[\dfrac{2\pi x}{3}\right]}{2\pi^3}$$

14: FOURIER $(x^3, x, -3, 3, 4)$

15: $$-\frac{27\,(8\pi^2 - 3)\,\sin\left[\dfrac{4\pi x}{3}\right]}{16\pi^3} + \frac{6\,(3\pi^2 - 2)\,\sin(\pi x)}{\pi^3}$$

$$-\frac{27\,(2\pi^2 - 3)\,\sin\left[\dfrac{2\pi x}{3}\right]}{2\pi^3} + \frac{54\,(\pi^2 - 6)\,\sin\left[\dfrac{\pi x}{3}\right]}{\pi^3}$$

16: FOURIER $(x^3, x, -3, 3, 6)$

17: $-\dfrac{3\,(6\,\pi^2 - 1)\,\text{SIN}\,(2\,\pi\,x)}{2\,\pi^3} + \dfrac{54\,(25\,\pi^2 - 6)\,\text{SIN}\left[\dfrac{5\,\pi\,x}{3}\right]}{125\,\pi^3}$

$-\dfrac{27\,(8\,\pi^2 - 3)\,\text{SIN}\left[\dfrac{4\,\pi\,x}{3}\right]}{16\,\pi^3} + \dfrac{3\,(3\,\pi^2 - 2)\,\text{SIN}\,(\pi\,x)}{\pi^3}$

$-\dfrac{27\,(2\,\pi^2 - 3)\,\text{SIN}\left[\dfrac{2\,\pi\,x}{3}\right]}{2\,\pi^3} + \dfrac{54\,(\pi^2 - 6)\,\text{SIN}\left[\dfrac{\pi\,x}{3}\right]}{\pi^3}$

The additional lines in the output are the part of the expression hidden off the right side of the screen. Of course, this part of an expression can be accessed through careful use of the direction keys to move the highlight to the right through the expression. The direction keys are used to control the level of the highlight. The ↓ and ↑ keys change the level in the expression, and the → and ← keys move the highlight in the appropriate direction at the current level of the highlight.

In order to see the behavior as n changes, we need to plot these approximations and the original function. There are two ways to do this efficiently. One way is to put the four functions as separate components into a vector and plot the entire vector at once with the Plot Overlay Plot command. The other way is to divide the screen into two windows and plot the functions one at a time. This can be done with the Window Split command or with Plot Under or Plot Beside. We'll show and explain the use of the Window Split command. Try the other ways on your own.

Execute the Window Split command. Make the horizontal split at line 15. Then Window Designate the upper window for 2-D Plot. Now as Plot and Algebra commands are issued, Derive automatically will move back and forth between the plotting window and the algebra window. Another way to move back and forth between the two windows is with the Windows

Next or **Windows Previous** commands. Derive indicates the active window by
background shading around the number label of the active window.

Move to the algebra window by selecting **Algebra**, and use the direction
keys (↑ or ↓) to highlight the first function ($n = 2$). Move back to the plotting
window and plot this function through the **Plot Plot** command. The result
is the following display:

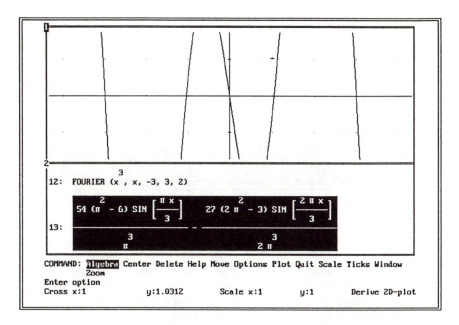

Not enough of the function is shown to determine its global behavior. Try
changing the scale using the **Scale** command. Set the x-scale to 3 and the
y-scale to 15. Now, reissue the **Plot** command. This plot is much better. It
shows the periodic behavior of the function and its maximum and minimum
values.

Now, it is only a matter of moving back and forth between the two
windows (using **Plot** and **Algebra**), highlighting the next function, and
plotting it on the same axes as the previous plots. On a color monitor, each

plot is done in a different color. The resulting black-and-white plot of all four functions is as shown.

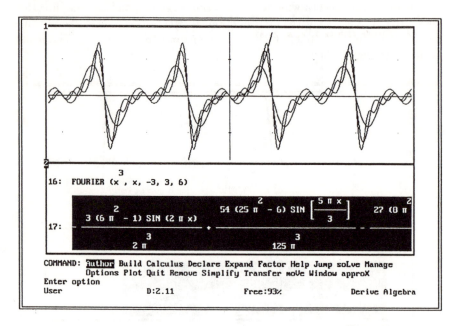

As the plotting takes place, notice the improvement in the series approximations as n increases. But at the same time, the oscillations don't disappear. This is a phenomenon discussed in many differential-equations textbooks.

Once this plot window is no longer needed, close the plotting window by executing the Window Close command.

Problem B: Let's try to find the Fourier series approximation to another function, $u(x) = \sin(x^2)$, which will result in some tougher integration to find the components of the series.

Solution to B: For $u(x) = \sin(x^2)$, the command

```
FOURIER(sin(x^2),x,0,2,4)
```

sets up the integration for the series approximation to four terms of $u(x)$ in the region $0 < x < 2$. Simplify performs the integration. The results are displayed in the following figure:

18: FOURIER (SIN (x²), x, 0, 2, 4)

$$19: \int_0^2 SIN(x^2) COS(4\pi x)\,dx\; COS(4\pi x) + \int_0^2 SIN(x^2) SIN(4\pi x)\,dx\; SIN($$

Since integral set-ups remain in the simplified expression, this means Derive is unable to perform the integration in the Exact mode. Change the mode to Mixed or Approximate through the Options Precision menu commands. Now re-execute the Simplify command on the FOURIER command line, or the integral results.

This time, numerical quadrature is used to perform the definite integration. The system will take quite a long time and may beep and issue the warning "Dubious Accuracy." This means the numerical integration algorithm is having difficulties satisfying the error tolerance set in the Digits mode of the Precision command. Let the algorithm continue its work. The display with the result extended on extra lines is shown.

$$20: \quad -\frac{303}{15643} COS\left[-\frac{1420}{113}x\right] + \frac{4065}{62998} SIN\left[\frac{1420}{113}x\right] - \frac{510}{12961} COS\left[\frac{1065}{113}x\right]$$

$$+ \frac{714}{7897} SIN\left[\frac{1065}{113}x\right] - \frac{2159}{17338} COS\left[\frac{710}{113}x\right] + \frac{8777}{62967} SIN\left[\frac{710}{113}x\right]$$

$$\frac{4483}{8355} COS\left[\frac{355}{113}x\right] - \frac{79}{385} SIN\left[\frac{355}{113}x\right] + \frac{8933}{22200}$$

Now, in light of the dubious accuracy, we should compare the plot of the approximation with that of the actual $u(x)$ function. Unfortunately, there is no way to do this efficiently. If the expressions are put in a vector as the two components, Derive thinks the expression is in polar form. So first Plot $\sin(x^2)$, and then return to the Algebra window and highlight the Fourier approximation. Return to the Plot window and issue the Plot command to plot the second function. These plots using the default scale parameters are as follows:

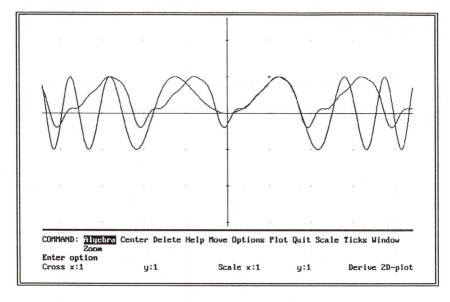

Even though the plot of the Fourier approximation looks strange, it matches the function quite well in the specified interval, $0 < x < 2$. The graph exhibits the periodic property of the Fourier series on the rest of the plot interval.

The utility file doesn't have commands for the Fourier sine series or the Fourier cosine series. If these expansions are needed, the actual definition for the coefficients for these series could be entered through the **Author** command. The easiest way to input the definitions for these two series is probably to modify the definition of the existing **FOURIER** command using the direction arrows, the highlight, the function keys, and the special control keys discussed in Section 1.2. For instance, a double sine series for $f(x, y)$ involves the following integration to determine the coefficients:

$$A_{mn} = \frac{4}{bc} \int_0^c \int_0^b f(x, y) \sin\left(\frac{m\pi}{b}x\right) \sin\left(\frac{n\pi}{c}y\right) dx dy.$$

If $f(x, y) = xy$, $0 < x < 1$, and $0 < y < 1$, then the Derive in-line command for this integration is

INT(INT(xy sin(m pi x/b) sin(n pi y/c),x,0,b),y,0,c) .

When the **Precision** is still in the **Approximate** or **Mixed** mode, this operation is performed through the **Author** of the command line and selection of

Simplify to obtain the following display with three digits of accuracy. Once again, the additional line under the expression for the approximation is the rest of the expression, usually hidden off the right side of the screen.

21: $\displaystyle\int_{\theta}^{c}\int_{\theta}^{b} x\, y\, \text{SIN}\left[\frac{m\,\pi\,x}{b}\right]\text{SIN}\left[\frac{n\,\pi\,y}{c}\right] dx\, dy$

23: $\displaystyle\left[\frac{0.101\, b^2\, c^2\, \text{COS}\,(3.14\, n)}{m\, n} - \frac{0.0322\, b^2\, c^2\, \text{SIN}\,(3.14\, n)}{m\, n^2}\right]\text{COS}\,(3.14\, m)$

$\displaystyle +\left[\frac{0.0102\, b^2\, c^2\, \text{SIN}\,(3.14\, n)}{m^2\, n^2} - \frac{0.0322\, b^2\, c^2\, \text{COS}\,(3.14\, n)}{m^2\, n}\right]\text{SIN}\,(3.14\, m)$

Exercise 7.2: Evaluating Fourier Series

There is in mathematics hardly a single infinite series of which the sum is determined in a rigorous way.

—Niels Henrik Abel [1826]

Subject: Approximating Functions Using Fourier Series

Purpose: To use Derive to find terms of the Fourier series for seven functions using the FOURIER function in the utility file INT_APPS.MTH.

References: Sections 1.12 and 7.1

Given: The following seven definitions of functions, $f_i(t)$, $i = 1, 2, \ldots, 7$:

$$f_1(x) = \begin{cases} 0, & \text{if } -\pi \le x < 0 \\ 1, & \text{if } 0 \le x \le \pi \end{cases},$$

$$f_2(x) = \begin{cases} -x, & \text{if } -2 \le x < 0 \\ x, & \text{if } 0 \le x \le 2 \end{cases},$$

$$f_3(x) = \begin{cases} 0, & \text{if } -3 \le x < -1 \\ 1, & \text{if } -1 \le x < 1 \\ 0, & \text{if } 1 \le x \le 3 \end{cases},$$

$$f_4(x) = \begin{cases} 0, & \text{if } -\pi \le x < 0 \\ x, & \text{if } 0 \le x \le \pi \end{cases},$$

$$f_5(x) = -x \qquad \text{if } -\pi < x \le 0,$$

$$f_6(x) = e^x \qquad \text{if } -\pi < x \le \pi,$$

$$f_7(x) = \cos(x^2)e^x \quad \text{if } -\pi < x \le \pi.$$

Exercises:

1. Load the utility file INT_APPS.MTH into the work space. Use the function **FOURIER** to attempt to find the first three terms of the Fourier series for these seven functions. Which functions can't it evaluate? If the evaluation cannot be performed in the **Exact** mode, change Derive to the **Approximate** mode. Do any of the calculations give a warning of "**Dubious Accuracy**"?

2. Plot each of the seven functions and its 3-term approximation in the interval $-8 < x < 8$. What are the periods of these functions? Which functions are even or odd?

3. Find the first six terms for each of these seven functions.

4. Plot the 3-term and the 6-term approximations for each of the functions on the interval $-8 < x < 8$. Does the approximation for each function improve with more terms?

Example 7.3: Separation of Variables

Education is not received. It is achieved.

—Anonymous

Subject: Separation of Variables for the Wave Equation

References: Sections 1.7, 1.10, and 1.13

Problem: The vertical displacement $u(x,t)$ of a vibrating string of length l is determined from solving the wave equation, a hyperbolic partial differential equation of the form

$$k^2 \frac{\partial^2 u}{\partial x^2} = \frac{\partial^2 u}{\partial t^2}, \quad 0 < x < l, \ t > 0,$$

where

$$u(0,t) = 0,$$
$$u(l,t) = 0,$$
$$u(x,0) = bx(l - x),$$
$$\frac{\partial u(x,0)}{\partial t} = c.$$

Here, x is the horizontal location, t is time, k is a parameter relating string tension and mass, b is a parameter controlling the amplitude of the initial displacement, and c is a parameter controlling the initial velocity of the string. The initial conditions given simulate plucking the string at its middle (at $x = l/2$).

Solve this equation and investigate its solution and standing waves at $t = 1$ for specified values of the parameters k, b, and c.

Solution: First, let's use Derive's plotting capability to visualize the initial geometry of the problem. Author the initial displacement into the work area by entering

$$\boxed{\texttt{bx(1 - x)}}.$$

Derive's 2-D plotter won't plot an expression like the given initial condition with extra unknown parameters such as b and l because they make it a higher-dimensional expression. For visualization only, substitute a value of

1 for b and 3 for l using the **Manage Substitute** command. Just type 1 for the b value, 3 for the l value, and **Enter** for the x value. Now, execute the **Plot Overlay Plot** command. The plot of the initial displacement of the string with these parameter values and using the default plotting parameter values is as shown.

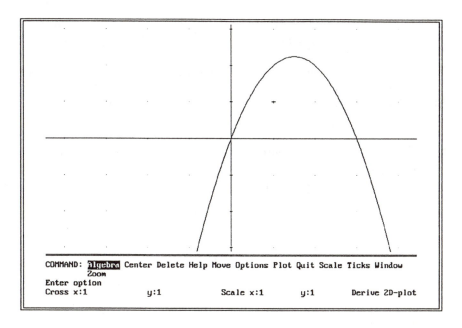

Derive plots the function over the entire domain of the x-axis shown in the window, even if a specific interval is established for x using the **Declare Variable** and **Interval** menu commands. The actual function describing the initial condition exists only between $x = 0$ and $x = l = 3$ (in this case).

The solution technique of separation of variables assumes that the solution $u(x, t)$ is a product of a function of x only $(X(x))$ and a function of t only $(T(t))$, so that $u(x, t) = X(x)T(t)$. Making this substitution, the equation becomes

$$k^2 X''T = XT''.$$

Separate the variables by dividing common terms and equating to a constant, w, to get

$$\frac{X''}{X} = \frac{T''}{k^2 T} = w.$$

We use w instead of the more common λ because λ is one of the Greek characters that Derive does not have available for use. This requires solving the two ordinary differential equations

$$X'' - wX = 0$$

and

$$T'' - wk^2 T = 0.$$

The boundary conditions for the ordinary differential equation in x come from the partial differential equation's boundary conditions, so for this equation, $X(0) = 0$ and $X(l) = 0$.

Commands from Derive's utility file ODE2.MTH can be used to solve for $X(x)$ and $T(t)$ for the three possibilities for w: $w > 0$, $w < 0$, and $w = 0$. The utility file is loaded into the work area with the **Transfer Load Utility** command.

For the situation where $w > 0$, the command **Declare Variable** is used to declare w **Positive**. Hereafter, we'll keep track of the sign of w as we go through the three cases. Now solving for X, **Author** and **Simplify** the command $\boxed{\text{DSOLVE2}(0,-w,0,x)}$. The result is

```
3:   DSOLVE2 (0, -w, 0, x)

              √w x          - √w x
4:   c1 ℮          + c2 ℮
```

The c_1 and c_2 are arbitrary constants.

For this case, boundary conditions of $c_1 + c_2 = 0$ and $c_1 e^{\sqrt{w}\,l} + c_2 e^{-\sqrt{w}\,l} = 0$ lead to $c_1 = 0$ and $c_2 = 0$ and the trivial solution.

The easiest way to change the sign on w to proceed with the second case is to leave w declared as positive and change the first equation to $X'' + wX = 0$. For this new equation, the screen output of the solution operations is shown.

```
5:   DSOLVE2 (0, w, 0, x)

6:   c1 COS (√w x) + c2 SIN (√w x)
```

Derive doesn't have a way to restrict variables, like n, to integer values. The two boundary conditions lead to $c_2 = 0$ and $\sin \sqrt{w}\, l = 0$. Therefore, $\sqrt{w} = n\pi/l$, $n = 1, 2, \ldots$. For this case of w, the equation for $T(t)$ becomes

$$T'' + \frac{n^2\pi^2}{l^2} k^2 T = 0.$$

The commands to solve this equation for T are similar to those for the equation in x. Before solving, we must use the Declare Variable command to declare n, k, and l to be Positive. The screen display of the solution commands is provided.

$$
7: \quad \text{DSOLVE2} \left[0, \ \frac{n^2 \pi^2 k^2}{l^2}, \ 0, \ t \right]
$$

$$
8: \quad c1 \ \text{COS} \left[\frac{\pi k n t}{l} \right] + c2 \ \text{SIN} \left[\frac{\pi k n t}{l} \right]
$$

For the last case of $w = 0$, the equation is simply $X'' = 0$, which has the solution $c_1 + c_2 x$. The boundary conditions force the trivial solution $(c_1 = c_2 = 0)$ for this case.

Therefore, the solutions for $u(x,t)$ have the form

$$
\left[\left(a(n) \cos \frac{n\pi k}{l} t + b(n) \sin \frac{n\pi k}{l} t \right) \right] \sin \frac{n\pi}{l} x.
$$

By the superposition principle, we construct the solution as

$$
u(x,t) = \sum_{n=1}^{\infty} \left[\left(a(n) \cos \frac{n\pi k}{l} t + b(n) \sin \frac{n\pi k}{l} t \right) \right] \sin \frac{n\pi}{l} x.
$$

For this solution, the initial conditions for the partial differential equation are

$$
u(x,0) = \sum_{n=1}^{\infty} a(n) \sin \frac{n\pi}{l} x = bx(l - x)
$$

and

$$
\frac{\partial u(x,0)}{\partial t} = \sum_{n=1}^{\infty} b(n) \frac{n\pi k}{l} \sin \frac{n\pi}{l} x = c.
$$

In this formula, the $b(n)$ are coefficients and are not the same as parameter b. The coefficients, $a(n)$ and $b(n)$, are evaluated as Fourier coefficients by

$$a(n) = \int_0^l \frac{2}{l} bx(l-x) \sin \frac{n\pi}{l} x \, dx$$

and

$$b(n) = \int_0^l \frac{2}{n\pi k} c \sin \frac{n\pi}{l} x \, dx \quad .$$

To evaluate these integrals, **Author** the expression for the integrand `2/l(bx(l-x)sin(n pi x/l))` and select the `Calculus Integrate` menu command. Input the lower and upper limits, 0 and l, and `Simplify`. (Remember, the **Tab** key moves the cursor through the various parts of the menu.) The resulting display and expression for finding $a(n)$ are as follows:

9: $\displaystyle \frac{2}{l} \left[b\ x\ (l - x)\ \mathrm{SIN}\left[\frac{n\ \pi\ x}{l}\right]\right]$

10: $\displaystyle \int_0^l \frac{2}{l} \left[b\ x\ (l - x)\ \mathrm{SIN}\left[\frac{n\ \pi\ x}{l}\right]\right] dx$

11: $\displaystyle -\frac{4\ b\ l^2\ \mathrm{COS}\ (\pi\ n)}{\pi^3\ n^3} - \frac{2\ b\ l^2\ \mathrm{SIN}\ (\pi\ n)}{\pi^2\ n^2} + \frac{4\ b\ l^2}{\pi^3\ n^3}$

An alternate procedure for performing integration is to use the in-line command `INT`. The command for $b(n)$ is

`INT(2c/(n pi k) sin(n pi x/l),x,0,l)` .

Be careful to notice the difference between the l and 1 in the input string. Select the command `Simplify` to perform this operation. The following screen display shows the results of these operations:

$$
12: \quad \int_{0}^{1} \frac{2\,c}{n\,\pi\,k}\,\text{SIN}\left[\frac{n\,\pi\,x}{1}\right]\,dx
$$

$$
13: \quad \frac{2\,c\,1}{\pi^{2}\,k\,n^{2}} - \frac{2\,c\,1\,\text{COS}\,(\pi\,n)}{\pi^{2}\,k\,n^{2}}
$$

Since Derive still doesn't know that n is an integer, it doesn't simplify $\sin(n\pi)$ to 0 or $\cos(n\pi)$ to $(-1)^{n-1}$. This is one of Derive's limitations. So with this type of manual simplification, the solution to our problem becomes

$u(x,t) =$

$$
\sum_{n=1}^{\infty}\left(\frac{4l^{2}b}{\pi^{3}n^{3}}(1-(-1)^{n})\cos\frac{n\pi k}{l}t + \frac{2cl}{n^{2}k\pi^{2}}(1+(-1)^{n})\sin\frac{n\pi k}{l}t\right)\sin\frac{n\pi}{l}x.
$$

Let's try plotting the first two approximate solutions ($m = 1$ and 2), where m is the upper limit of the index n in the summation, at $t = 1$ for the following specified values of the parameters: $b = 1.1$, $c = 1.5$, $k = 4.0$, and $l = 3$. Author the general expression for m terms of the series solution into the work area. Derive shows this as follows:

$$
16: \quad \sum_{n=1}^{M}\left[\frac{4\,1^{2}\,b}{\pi^{3}\,n^{3}}\,(1-(-1)^{n})\,\text{COS}\left[\frac{n\,\pi\,k\,t}{1}\right]\,+\right.
$$

$$
\left.\frac{2\,c\,1}{n^{2}\,k\,\pi^{2}}\,(1+(-1)^{n})\,\text{SIN}\left[\frac{n\,\pi\,k\,t}{1}\right]\right]\,\text{SIN}\left[\frac{n\,\pi\,x}{1}\right]
$$

Then Manage Substitute these variable and parameter values. For $m = 1$, the *Simplified* expression and its plot, centered at $(1.5, 0)$ with x-scale set to 0.5 and y-scale set to 2, are shown. Remember, we are only interested in the region $0 < x < 3$. The resulting summation expression is shown. The additional lines are the continuation of the expression that is off the right of the screen in the Derive display.

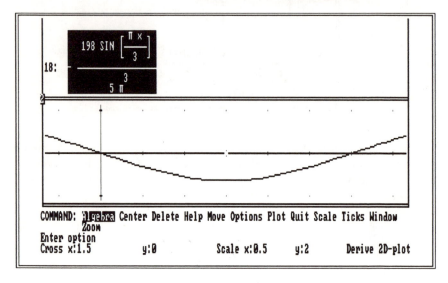

The truncated series approximation with $m = 2$ and the plot of this expression (shown with the previous plot for the expression with $m = 1$) are shown in the Derive output screen as follows:

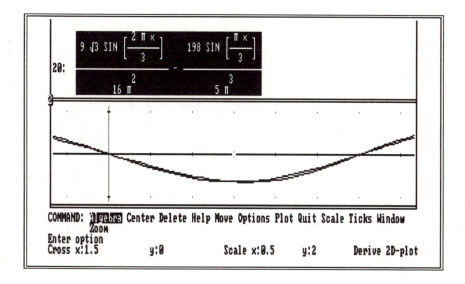

Various parameter studies can be done by varying the values of each of the parameters (b, c, k, and l). An example of this might be to determine the value of c that restricts the amplitude to values less than 0.2. Go ahead and try some of these for this problem by substituting different values for these parameters and plotting the resulting solutions.

Now, let's look at a 4-term approximation to the series solution at various values of t ($t = 1$, 2, 3, and 3.7). Let's keep the previously specified values of the parameters: $b = 1.1$, $c = 1.5$, $k = 4.0$, and $l = 3$. Using the general expression for the solution, **Manage Substitute** these parameter values (with $m = 4$ and keeping x and t as unknown variables for now). **Simplify** this expression to obtain the 4-term approximation to $u(x, t)$. This expression is shown, with the extra terms usually off the screen to the right wrapped around below the beginning of the expression.

$$18: \quad \frac{9 \sin\left[\frac{16 \pi t}{3}\right] \sin\left[\frac{4 \pi x}{3}\right]}{32 \pi^2} + \frac{44 \cos(4 \pi t) \sin(\pi x)}{15 \pi^3} +$$

$$\frac{9 \sin\left[\frac{8 \pi t}{3}\right] \sin\left[\frac{2 \pi x}{3}\right]}{8 \pi^2} + \frac{396 \cos\left[\frac{4 \pi t}{3}\right] \sin\left[\frac{\pi x}{3}\right]}{5 \pi^3}$$

Use the Manage Substitute command to input the four values of t, one at a time, into this function. Simplify each function and put them together as components of a vector. The easiest way to do this is to Author the expression [#a,#b,#c,#d] where a, b, c, and d are the Derive statement numbers in the work area of the four functions. Plot all four functions at once on the same axes using the Plot Plot menu command. The resulting plots are as shown.

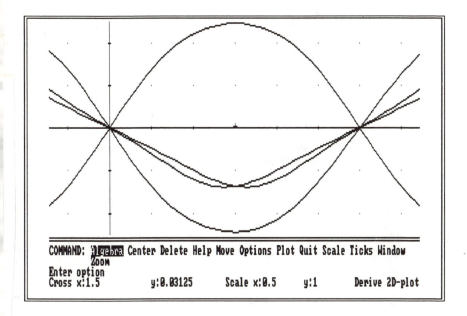

```
COMMAND: Algebra Center Delete Help Move Options Plot Quit Scale Ticks Window
         Zoom
Enter option
Cross x:1.5          y:0.03125       Scale x:0.5      y:1         Derive 2D-plot
```

Exercise 7.4: Heat Equation

A traveler who refuses to pass over a bridge until he has personally tested the soundness of every part of it is not likely to go far; something must be risked, even in mathematics.

—Horace Lamb

Subject: Separation of Variables for the Heat Equation

Purpose: To solve the heat equation with nonhomogeneous boundary conditions and to investigate the effects of varying parameters in the initial and boundary conditions.

References: Sections 1.10, 7.1, and 7.3

Given: The temperature in degrees $u(x,t)$ of a rod of length 15 meters is modeled by the following heat equation:

$$ku_{xx} = u_t, \quad 0 < x < 15, \quad t > 0,$$

where

$$\begin{aligned} u(0,t) &= a, \\ u(15,t) &= 0, \\ u(x,0) &= b, \end{aligned}$$

k is a constant thermal diffusivity, and a and b are parameter values for temperature.

Exercises:

1. Use the change of dependent variable $(u(x,t) = v(x,t) + w(x))$ to transform the given equation into a heat equation for v with homogeneous boundary conditions and into a second-order ordinary differential equation for w with nonhomogeneous boundary conditions. The two parts of the solution v and w are called the transient and steady-state solutions, respectively.

2. Solve this heat equation for v and the differential equation for w. Does the steady-state solution depend on k, a, or b? Explain why.

3. Find the expression for the temperature distribution $u(x,t)$.

4. If $k = 0.853$ (assume this value for k for the rest of the problem), use the first term in the series for the transient solution and the steady-state solution to approximate the temperature at $x = 6$ when $t = 1$, 10, 20, and 40.

5. What are the values computed in #4 when $a = 100$ and $b = 100$? Use 2 terms of the series to approximate the same values.

6. If $a = 100$ and $b = 100$, plot on the same axes the 3-term approximations of the temperature distributions for $t = 0$, 10, and 20 and the steady-state solution. How close is the temperature distribution to steady state by $t = 20$?

7. If $a = 100$, plot on the same axes the 3-term approximations of the temperature distribution for $t = 5$ when $b = 0$, 10, 20, and 40. Does the initial temperature have a lasting effect on the temperature distribution?

8. If $b = 100$, plot on the same axes the 3-term approximation of the temperature distribution for $t = 5$ when $a = 0$, 10, 20, and 40. Does the value of the boundary condition parameter a have a lasting effect on the temperature distribution?

9. Ask yourself other questions related to this problem that you would like to investigate. Determine if Derive is a proper tool to use in this investigation by trying to answer your questions.

Other Reading

I cannot live without books.

—Thomas Jefferson [1815]

The following references, along with the *Derive User Manual*, provide more information about Derive or other computer algebra systems:

Arney, David C., *Derive Laboratory Manual for Differential Equations*, Reading, MA: Addison-Wesley, 1991.

Arney, David C., *Exploring Calculus with Derive*, Reading, MA: Addison-Wesley, 1992.

Arney, David C., *Instructor's Manual to Accompany the Student Edition of Derive*, Reading, MA: Addison-Wesley, 1992.

Arney, David C., *The Student Edition of Derive Users Manual*, Reading, MA: Addison-Wesley, 1992.

Braden, B., Krug, D.K., McCartney, P.N., and Wilkinson, S., *Discovering Calculus with Mathematica*, New York: John Wiley, 1992.

Demana, Franklin; Waits, Bert K.; and Harvey, John; Editors, *Proceedings of the Second Annual Conference on Technology in Collegiate Mathematics*, Reading, MA: Addison-Wesley, 1990. This volume has numerous articles on the use of computer algebra systems in calculus courses.

Ellis, Wade, Jr. and Lodi, Ed, *Maple for the Calculus Student: A Tutorial*, Pacific Grove, CA: Brooks/Cole, 1989.

Ellis, Wade, Jr. and Lodi, Ed, *A Tutorial Introduction to Derive*, Pacific Grove, CA: Brooks/Cole, 1991.

Evans, Benny and Johnson, Jerry, *Uses of Technology in the Mathematics Curriculum*, Stillwater, OK: Cipher Systems, 1990.

Fattahi, Abi, *Maple V, Calculus Labs*, Pacific Grove, CA: Brooks/Cole, 1992.

Geddes, K.O., Marshman, B.J., McGee, I.J., Ponzo, P.J., and Char, B.W., *Maple, Calculus Workbook*, Univ. of Waterloo, 1988.

Gilligan, Lawrence G. and Marquardt, James F., Sr., *Calculus and the Derive Program: Experiments with the Computer*, Cincinnati: Gilmar, 1990. Write to Gilmar Publishing, P.O. Box 6376, Cincinnati, OH 45206.

Glynn, Jerry, *Exploring Math from Algebra to Calculus with Derive, A Mathematical Assistant*, Urbana, IL: MathWare, 1989. Write to MathWare, 604 E. Mumford Drive, Urbana, IL 61801.

Kerchkove, Michael G. and Hall, Van C., *Calculus Laboratories with Mathematica*, Vol. 1, New York: McGraw-Hill, 1993.

Leinbach, L. Carl, *Calculus Laboratories Using Derive*, Belmont, CA: Wadsworth, 1991.

Maeder, Roman, *Programming in Mathematica*, Reading, MA: Addison-Wesley, 1991.

Olwell, David H. and Driscoll, Patrick J., *Calculus and Derive*, New York: Saunders, 1992.

Page, W., "Computer Algebra Systems: Issues and Inquiries," *Computers and Mathematics Applications*, Vol. 19, 1990, pp. 51–69.

Porta, H., Uhl, J.J., and Brown, D., *Calculus and Mathematica*, Reading, MA: Addison-Wesley, 1991.

Salter, Mary and Gilligan, Lawrence, *Linear Algebra Experiments Using the Derive Program*, copyrighted by Mary Salter and Lawrence Gilligan, 1992.

Small, Donald B. and Hosack, John M., *Explorations in Calculus with a Computer Algebra System*, New York: McGraw-Hill, 1990.

Small, Donald B., Hosack, John M., and Lane, K., "Computer Algebra Systems in Undergraduate Instruction," *College Mathematics Journal*, Vol. 17, November 1986, p. 423.

Wolfram, S., *Mathematica—A System for Doing Mathematics by Computer*, Reading, MA: Addison-Wesley, 1988.

Zorn, P., "Computer Symbolic Manipulation in Elementary Calculus," *The Future of College Mathematics*, New York: Springer, 1983, pp. 237–249.

Index

One writer, for instance, excels at a plan or a title page, another works away the body of the book, and a third is a dab at an index.

—Oliver Goldsmith, *The Bee* [1759]

Where the statue stood
Of Newton with his prism and silent face,
The marble index of a mind forever
Voyaging through strange seas of thought,
alone.

—William Wordsworth, *The Prelude*, written [1799–1805]

Here the people could stand it no longer and complained of the long voyage; but the Admiral cheered them as best he could, holding out good hope of the advantages they would have. He added that it was useless to complain, he had come [to go] to the Indies, and so had to continue it until he found them, with the help of Our Lord.

—Christopher Columbus [October 10, 1492]